# Wiring and Lighting

# WIRING
# & LIGHTING

MIKE LAWRENCE

The Crowood Press

First published in 1991 by
The Crowood Press Ltd
Ramsbury, Marlborough
Wiltshire SN8 2HR

**www.crowood.com**

This impression 2005

British Library Cataloguing in Publication Data

Lawrence, Mike *1947–*
    Wiring and lighting.
    1. Electric wiring systems
    I. Title
    621.31924

ISBN 1 85223 514 4


Acknowledgements

Line-drawings by Andrew Green.

The author would like to thank the following companies and organizations for
providing the photographs listed below:

Robert Harding Picture Library (pages 7, 10, 12, 13, 14, 16, 17, 18);
Marshall Cavendish (pages 35, 41, 42, 45, 47, 55, 61, 63, 65, 71, 72, 83, 85);
Ceiling Master (page 39).

Publisher's Note

Whilst the information in this book was correct at the time of going to press, the reader is advised
always to consult the Electricity Board's current wiring regulations before undertaking any
electrical work.

Typeset by Acūté, Stroud, Glos.
Printed and bound by Times Offset (M) Sdn Bhd, Malaysia.

# Contents

# Introduction

Of all the various activities that do-it-yourselfers get up to, wiring work requires the least skill. It certainly needs greater technical knowledge, more care and perhaps greater patience than other activities like plastering or paper hanging, but there is no great difficulty in doing wiring work well.

Unfortunately, this fact does not necessarily make people more confident about tackling jobs related to electricity. There is a certain mystique about dealing with something you cannot see, and many people will sympathize with the old person who, so the story goes, was afraid to change a light bulb in case the current ran out of the lamp-holder. Yet the analogy with plumbing is apt; good wiring practice is all about avoiding leaks, and it is much easier to make a 'leak-proof' electrical connection than to guarantee a perfect plumbing joint.

The secret of success – and of conquering your fears about electrical work – is simple: master the techniques first of all, and then have confidence in your workmanship. This book will help you with the first and encourage you with the second.

## IMPORTANT NEWS FOR DIY ELECTRICIANS
### DIY WIRING WORK AND THE BUILDING REGULATIONS
With effect from January 2005, all new domestic wiring work in England and Wales must comply with the requirements of a recently introduced section of the Building Regulations. Part P, entitled Electrical Safety, covers the design, installation, inspection and testing of new wiring work. It does not apply in Scotland, where electrical installation work has been subject to Scottish Building Regulations control since 1969.

### GETTING APPROVAL
DIY wiring work is still permitted following the introduction of Part P. However, to ensure that it complies with the requirements of Part P, you must have your work inspected and tested on completion by an electrician who will issue a Minor Electrical Installation Works Certificate if it is satisfactory. Depending on the scale of the work involved, you may also have to notify your local authority Building Control Department. This involves submitting a building notice to the local authority before you start work, and paying a building control fee to have the work inspected and tested when you have completed it.

### EXEMPT JOBS
Some minor jobs do not need to be notified to building control. These include:
- Replacing existing wiring accessories, such as switches, socket outlets and ceiling roses, and replacing their mounting boxes;
- Replacing circuit cable that has been damaged by, for example, impact, fire or rodent attack;
- Adding new lighting points and switches to an existing lighting circuit, except in a kitchen or bathroom;
- Adding new socket outlets or fused spurs to an existing ring main or radial circuit, except in a kitchen or bathroom.

### JOBS THAT MUST BE NOTIFIED
- The provision of any new indoor circuit, such as one supplying an electric shower.
- Any new wiring work in a kitchen or a bathroom (defined as any location containing a bath or shower). This includes the installation of electric underfloor heating. However, replacement work (see above) is exempt in both these rooms.
- The installation of extra-low-voltage lighting. However, installing pre-assembled extra-low-voltage light fittings that have CE approval is exempt.
- The installation of new outdoor lighting and power circuits.

If you are in any doubt as to whether the DIY wiring work you plan to do requires notification, contact your local authority Building Control Office for advice.

### CABLE CORES
With effect from 31 March 2006, you must use cable with cores insulated in new colours for all new fixed wiring work. This change has been introduced as part of a process of product harmonization across the European Union, some 35 years after flex core colours were changed from red and black to brown and blue.

- Two-core-and-earth cable will have the live (phase) core coloured brown, and the neutral core coloured blue, instead of the red and black currently in use. The earth core will remain a bare copper conductor. As before, you must cover it in green-and-yellow PVC sleeving wherever it is exposed at terminal connections.
- Three-core-and-earth cable (used in the home only for wiring two-way switching arrangements) will have cores colour-coded brown, black and grey instead of red, yellow and blue. The earth core will again be a bare copper conductor (see above).

Cable with the new core colours is already available, and you can use it now on any new wiring work you carry out. However, new installations or alterations to existing installations may use either the new or old colours, but not both. When you have used it to alter or extend an existing wiring installation, you must place a warning notice to this effect at the fuse board or consumer unit. Its wording is as follows:

CAUTION
This installation has wiring colours
to two versions of BS7671. Great care should be taken before undertaking extension, alteration or repair that all conductors are correctly identified.

The illustrations in Wiring & Lighting show new work being carried out using two-core-and-earth cable with red and black cores. When using cable with the new core colours, remember that new brown = old red and new blue = old black.

# THE BASICS

Wiring work is surprisingly straightforward, but it can be somewhat repetitive; whatever the project, you wind up doing the same relatively simple jobs over and over again. However, that does not make the work any less rewarding, because the end product can yield all sorts of benefits. For example, new lighting effects can improve your interior decor dramatically, while extra power supplies mean you can enjoy the use of a wider range of labour-saving electrical equipment and leisure-time appliances.

To become a good do-it-yourself electrician, you need to understand how your wiring system works, familiarize yourself with the raw materials you will be using, master the basic techniques and decide what jobs you can (or want to) undertake. This chapter takes you through all these steps one by one.

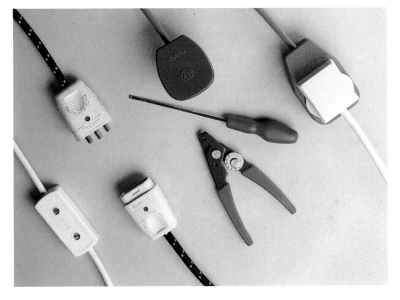

## How Electricity Works

To understand your house wiring system, it helps to have an idea of how electricity works. The best analogy is with water flowing through a pipe. Both electricity and water can perform 'work'; for example, water can turn a water-wheel and drive a machine, while electricity can produce light if it flows through a lamp, heat if it flows through a fire element, or rotation in an electric motor.

What drives both water and electricity is a difference in pressure between the ends of the circuit through which it flows; the greater the pressure, the greater the flow. Acting against this pressure is a factor called *impedance*. Going back to our water analogy again, it is clearly easier for water to flow through a wide pipe than through a narrow one. In electrical terms impedance divides materials into two groups – *conductors* and *insulators*. Conductors have low impedance to the flow of electricity, while insulators have high impedance – to the point of not letting electricity flow at all.

Just like water, electricity needs a circuit to flow round if it is to perform work.

The circuit starts at your local supply substation, and the current flows through a conductor (the *live* wire) to wherever it is needed. It then returns along another conductor (the *neutral* wire) back to its source. The flow of current (measured in units called *amperes* – amps or A for short) is driven through the circuit by the pressure (or potential) difference measured in *volts* (V), acting against the resistance of the circuit (measured in ohms – Ω for short). The work it does is measured in watts (W) – *see* Useful Formulae.

In your home, cables deliver electricity to wherever it is needed around the house, and then return it to the system. Each cable contains live and neutral conductors, plus a third conductor called the earth continuity conductor, or *earth* for short.

Just like a plumbing leak, electricity can escape and flow to earth on its rounds. If you touch a live part of the system, the current flows through your body to the ground, giving you a shock which can be fatal. To guard against this, everything metallic in your home's wiring system is connected to earth via the earth conductor in the circuit cables to provide a safe discharge path for any 'leaking' current.

Fig 1 (*above*) Start with the simplest jobs, and you will soon be able to progress to more complex projects.

---

**USEFUL FORMULAE**
Current flow, potential (pressure) difference and impedance are related by the equation:

volts = amps x ohms.

The amount of electricity consumed whenever work is done (lighting a lamp or driving a motor, for example) is measured in watts – the product of the supply voltage (240V in most of the UK) and the current drawn:

watts = volts × amps.

---

# Wiring around the House

So much for the theory: now let's follow the current flow round the wiring installation in a typical home, and see how everything fits together in real life.

## The Heart of the System

Your electricity supply enters the house via a thick supply cable, either run overhead from poles along the roadside, or more likely buried underground.

Within the house, this cable terminates at a sealed block called the *service head*. This contains the *service fuse*, which is there to prevent you demanding more current than the supply cable can safely carry without overheating. In most homes the supply cable can deliver 60 or 80 amps, but some homes with high electricity demand may have a 100-amp supply. If you try to draw more current than the cable can carry, the fuse melts and cuts off the power supply.

Two thick single-core cables, one with a red sheath and one black, emerge from the service head and run to your electricity meter, which records the amount of electricity you use in units called *kilowatt-hours* – kWh for short. A kilowatt is 1,000 watts, and one kilowatt-hour is the amount of electricity consumed by, for example, a 100-watt lamp burning for ten hours. The service head and the meter are both the property of your electricity distribution company and must not be tampered with under any circumstances.

From your meter, two more thick cables called *meter tails* run to the heart of your home's wiring system – what most people call the fuse-box. In a modern home, this will be a one-piece enclosure called a *consumer unit* that contains a main on/off switch, allowing you to cut off your entire system from the mains supply. It also controls the distribution of electricity to the various individual circuits in your home – to lights, socket outlets and to appliances like the cooker and immersion heater. In the older home, the main switch may have its own enclosure, and cables will run on to one or more separate fuse-boxes controlling the house circuits. Each of these may also have its own on/off switch.

Each of these circuits is protected by a circuit fuse, which does the same job of guarding that particular circuit against overloading as the service fuse does for

the supply cable. In older homes, this fuse is either a length of fuse wire or a small sealed cartridge fuse contained within a plug-in *fuse-holder*. In modern installations, the fuses are replaced by small electromechanical switches called *miniature circuit breakers* or MCBs. Each fuse or MCB is rated to match the current demand of the circuit it protects. Lighting circuits generally have a 5-amp fuse, power circuits a 20-amp or 30-amp fuse, while circuits to heavy current users such as cookers or instantaneous showers are protected by a 45-amp fuse. Fuse-holders are often colour-coded (*see* Fuse Colour Codes), while MCBs are simply labelled with their current rating.

Modern installations with MCBs in the consumer unit may also have a component called a *residual current device* (RCD) included in place of the traditional on/off switch or housed in a separate enclosure between the consumer unit and the meter. This is an invaluable safety device which detects any imbalance in the flow of electricity round your system, and cuts off the current in a fraction of a second if a fault occurs – fast enough to save your life if you touch something live, and on duty twenty-four hours a day to monitor the system against things like current leakage through faulty insulation, which could easily start a fire.

There is one last feature which you may find at your fuse-board – a supply of night-rate electricity via a separate meter and fuse-box. This feeds night storage heaters with electricity at a cheaper-than-usual rate, and may supply your immersion heater too.

From your fuse-board you will be able to see individual circuit cables running up towards the ceiling and down to the floor, ready to start their journey round the house. You should also be able to see a green or green-and-yellow cable running from your consumer unit or fuse-box to either a clamp on the sheath of the main supply cable, or else to a point underneath the floor where it connects to a metal rod driven into the ground. This is your system's main earthing point, and you should also find separate earth cables running to clamps on nearby gas and water supply pipes. These are known as *cross-bonding cables,* and are an essential safety feature of your wiring installation. For more details, *see* page 33.

For more details, *see* page 33.

---

**TIP 1**
Label your fuse-box so you can see at a glance which fuse-holder or MCB controls which circuit. In a typical system you are likely to have the following circuits;

- upstairs lighting (5-amp fuse)
- downstairs lighting (5-amp fuse)
- immersion heater (15-amp fuse)
- upstairs power (30-amp fuse)
- downstairs power (30-amp fuse)
- cooker (45-amp fuse)

---

**FUSE SIZES**
Several manufacturers of consumer units now supply MCBs rated at 6, 16 and 32 amps. These are alternatives to the 5, 15 and 30-amp ratings referred to here and elsewhere in this book.

---

**FUSE COLOUR CODES**
Individual fuse-holders or MCBs in your fuse-box or consumer unit may be colour-coded for ease of identification. The colour used denotes the circuit's current rating:

white = 5 amps
blue = 15 amps
yellow = 20 amps
red = 30 amps
green = 45 amps

Cartridge circuit fuses use the same colour coding, and apart from 15-amp and 20-amp fuses each is a different size, making it impossible to fit the wrong fuse in the fuse-holder.

# Wiring around the House

## Lighting Circuits

The thinnest cables emerging from your consumer unit or fuse-box supply your home's lighting circuits. Each is protected by a 5-amp circuit fuse or MCB, and most homes have two circuits – one supplying lights upstairs, the other downstairs. From the fuse-box, the cables are run in the void between the ground and first floors to supply the downstairs lights, and in the loft of a typical two-storey house to feed the upstairs lights. The cable runs on from one lighting point to the next, terminating at the most remote fitting. At each lighting point, a cable is connected in and runs down to the switch controlling that particular light.

It is a good idea for future reference to check out precisely which lights are on which circuit, and to record this information in a small notebook kept by the fuse-board. Do this by cutting off the power supply to one of the lighting circuits; turn off the main switch and remove the circuit fuse-holder if you have these, or if not, switch off the circuit MCB. Then go round the house and note which lights no longer work – they are on the circuit you have switched off. Repeat this for the other circuit(s). You may find lights that appear not to be on either circuit; these will take their supply from one of your power circuits, probably because they have been added to the system since the house was originally wired up. Note any that you find.

## Power Circuits

Thicker cables running from the consumer unit supply circuits to socket outlets around the house, and usually originate from a 30-amp fuse-holder or MCB in a modern installation. You will probably find two, or possibly three, separate circuits of this type. Again the cables generally run in the floor voids – below the ground floor for downstairs sockets, and between the ground and first floors for upstairs ones. Such circuits are usually wired up as a *ring circuit*, with the circuit cable running out to the sockets and back to the fuse-holder. The advantage of this is that current can flow to any socket either way round the ring, so it can effectively supply more sockets than if it was wired up like a lighting circuit.

In older homes, the power circuits *were* usually wired like lighting circuits, with the circuit terminating at the most remote socket. These circuits supplied sockets of different sizes, each with a different maximum current rating – 2 amps for plug-in lights, 5 amps for other small appliances and 15 amps for things like fires and kettles. If you still have such a system, it is long overdue for replacement.

As with your lighting circuits, check and record in your notebook which sockets are fed from which circuit.

## Other Circuits

Any remaining cables will supply individual circuits to appliances such as an electric cooker, an immersion heater in your hot water cylinder, or perhaps even an electric shower. In each case the cable runs direct from the fuse-holder or MCB to the appliance concerned only, and there is an on/off switch near the appliance that allows you to isolate it from the mains for maintenance or replacement.

The circuit to an immersion heater is usually rated at 15 amps, while those to cookers, showers and other large appliances may be rated at 30 or 45 amps, according to the amount of current each consumes.

In a very well-appointed system, you may also have additional circuits supplying outbuildings, outside lighting, even a burglar alarm system. You may find that not all of these are supplied from the main consumer unit or fuse-box; additional smaller units called *switch-fuse units* may have been added to the system at various times to increase its capacity.

> **TIP 2**
> If you have rewirable fuses in your fuse-box, make sure you keep a supply of fuse wire in various current ratings by the fuse-board in case you have to mend a fuse at any time.
>
> Better still, buy two spare fuse-holders and wire up one with 5-amp fuse wire and one with 30-amp wire. You will then be able to replace a blown fuse instantly, and re-wire it at your leisure with the power supply restored.
>
> If you have cartridge fuses, keep a supply of spare fuses of each current rating by the fuse-board.

# Raw Materials

Every wiring job you tackle will involve using electrical fittings of various types, plus cable and flex to connect things up. In this section you will find a brief description of the various types of wiring accessories available, and also information about cable, flex, light bulbs and tubes, plus some of the sundry other electrical bits and pieces you will need to complete the job.

## Power Circuit Accessories

Your home has two sorts of power circuits: one supplies socket outlets, allowing you to connnect portable appliances to the mains at convenient points around the house; the other supplies individual appliances such as cookers and immersion heaters.

**Socket outlets** These are by far the most numerous accessories in any home – a well-appointed system may have thirty or forty scattered around the house. All modern wiring installations have outlets with three rectangular holes in the faceplate, designed to accept fused plugs with rectangular pins. Each plug can supply up to 13 amps of current, enough for an appliance rated at up to 3kW – the typical current rating of an electric fire or fan heater, for example. In this case, the plug is fitted with a 13-amp fuse, which is colour coded brown for identification.

**Note** For more information on socket outlets, *see* pages 44 to 47, 59 and 68.

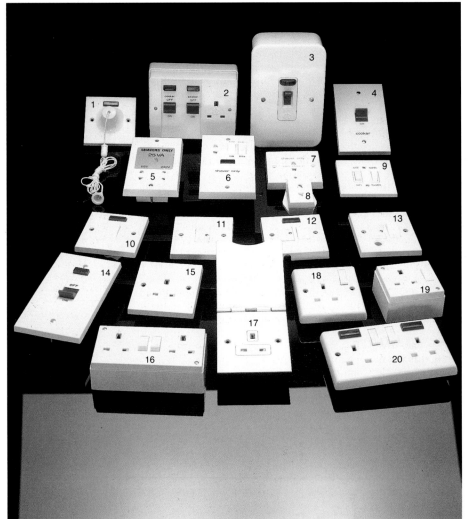

Fig 2 (*left*)
1  ceiling-mounted cord-pull DP switch
2  cooker control unit
3  45-amp surface-mounted DP switch
4  45-amp flush DP switch
5 and 6  shaver supply units with transformer
7 and 8  shaver socket and shaver adaptor
9  dual switch for immersion heaters
10  20-amp DP switch
11  fused connection unit
12  switched FCU with neon indicator and edge flex outlet
13  switched FCU with front flex outlet
14  45-amp DP switch with neon indicator
15  unswitched 13-amp single socket outlet
16  switched 13-amp double socket outlet on surface mounting box
17  unswitched 13-amp floor socket outlet
18  switched 13-amp single socket outlet
19  switched 13-amp single socket outlet on surface mounting box
20  switched 13-amp double socket outlet with neon indicators.

# Raw Materials

The same plug is also used to connect other smaller appliances to the mains, but if they are rated at less than 720 watts a 3-amp fuse is fitted in the plug.

Socket outlets are available with one, two or three sets of holes – known as single, double and triple sockets respectively. The most popular finish is white plastic, but you can also get socket outlets in various metallic finishes (brass and satin aluminium are the most popular of these) and in bright primary or soft pastel colours. The face-plate may have its own switch, and may also feature a neon indicator light to show whether the socket is switched on. They can be fitted on surface-mounted plastic or metal boxes (the latter are usually used in utility installations in places such as garages and workshops), or can be flush-mounted over galvanized steel boxes recessed into the wall surface.

You can also buy socket outlets for floor mounting – useful, for example, to avoid trailing flexes to a wall socket from an appliance on a piece of furniture in the middle of the room. Floor sockets have a spring-loaded cover plate to keep dust and dirt out of the contacts when the socket is not in use.

**Fused connection units** It is sometimes more convenient to connect certain appliances permanently to the mains, instead of plugging them into a socket outlet. Examples include freezers, where someone inadvertently pulling out the plug could ruin a lot of frozen food at great expense and other kitchen appliances such as washing machines, tumble driers and waste-disposal units that do not get moved around. A fused connection unit (FCU) provides the connection, with the appliance flex permanently connected to terminals within the accessory.

The FCU contains the same cartridge fuse as an ordinary fused plug; this fits in a removable fuse carrier slotted neatly into the unit's face-plate. As with a socket outlet, it may be switched or unswitched (the former is better for use with fixed appliances) and may have a neon indicator to show that the unit is on. The appliance flex may enter through a hole in the front of the face-plate, or in its edge.

FCUs are the same size as a single socket outlet, and come in the same range of finishes and may be flush or surface-mounted. If you want to mount two FCUs side by side, you can use special dual mounting boxes designed to accept two single face-plates side by side.

FCUs are also used to wire up branches (known as *spurs*) off main circuits, for example to supply a wall light via a power circuit. Since lighting circuits must be protected by a 5-amp fuse, and power circuits have a 30-amp fuse, the FCU is used to provide a sub-circuit to the light with the necessary level of protection. In this case a slightly different type of FCU is used – designed for a cable to be connected to its outlet terminals instead of an appliance flex. Again, switched or unswitched types are available, with or without neon indicators.

**Double-pole (DP) switches** These switches are used to control individual appliances that are permanently connected to the mains – cookers, showers or immersion heaters, for example. The term 'double-pole' means that the switch breaks both the live and neutral conductors in the circuit to the appliance, so enabling it to be completely isolated from the mains for maintenance or replacement. DP switches for wall mounting come in various current ratings, from 20 and 32 amps (both the same size as a single socket outlet), up to 45 and 60 (or 63) amps for large cookers. These need larger mounting boxes. The 20 amp size is available with a flex outlet – for things like immersion heaters and electric wall heaters – or with terminals for a cable connection. All are available with a neon indicator light if required to show that the switch is on.

Cooker switches are also available combined with a single 13-amp socket outlet – known as cooker control units. However, if the socket outlet is used the appliance flex could trail across a hot hotplate and become damaged by the heat, so it is generally safer to wire cookers to a plain DP switch instead, and to supply extra socket outlets elsewhere in the kitchen away from the cooker position.

DP switches are also available as cord-pull switches for ceiling mounting. This type is especially useful in bathrooms, where wall switches may not be fitted within reach of a bath or shower. They are available in two current ratings – 15 or

---

**Note** For more information on fused connection units, *see* pages 40, 49, 50 and 53. For information on double-pole switches *see* pages 53 to 57.

**CHECK**
- that the fuse fitted in an FCU is of the appropriate rating for the appliance it is serving. Use a 3-amp fuse for appliances rated at up to 720 watts, and 13-amp fuse for more powerful appliances
- that DP switches have a current rating to match the demand of the appliance they are controlling

**TIP**
You should aim for the following minimum number of *double* socket outlets in each room:

- living room 6
- dining room 3
- kitchen 4
- hall 1
- landing 1
- double bedroom 4
- single bedroom 3

**Note** No socket outlets are permitted in bathrooms.

# Raw Materials

16 amps and 40 or 45 amps – and usually have a neon indicator lamp on the switch base to show whether the switch is on. In addition, the higher-rated switches include a mechanical on/off indicator flag.

Lastly, a special dual switch is available for controlling dual-element or twin immersion heaters. One switch is an on/off switch; the other switches power one element or the other as required.

**Shaver sockets** Electric shavers can be run from ordinary socket outlets via a shaver adaptor, but it is preferable to provide special shaver sockets at convenient points such as in the bathroom or the bedroom. In bathrooms, a fitting called a shaver supply unit must be used for safety reasons; this contains a transformer that provides a power supply completely isolated from the mains, and which needs its own special deep mounting box. In other rooms a simple shaver socket, the same size as a single socket outlet, is all that is required.

## Lighting Circuit Accessories

Fewer different wiring accessories are needed for lighting circuits, since the system basically consists of light fittings and switches, plus a few components that are normally hidden from view.

**Light switches** Every light fitting needs an on/off switch, and the basic type has just one rocker switch on a square face-plate (the same size as a single socket outlet) which is known as a one-gang plate-switch. Switches with just two terminals are known as one-way switches, and control the light from just that one switch position. Switches with three terminals can be linked to other switches to provide control from two (or more) points, and are known as two-way switches.

Switches are available with one, two, three, four or even six rocker switches on a single face-plate (the four- and six-gang types are the same size as a double socket outlet). You can also buy narrow one-gang and two-gang switches called architrave switches, because they are designed to fit on narrow door architraves.

Wall switches come in a wide range of finishes, from white or coloured plastic to ornate brass. They can be flush- or surface-mounted. There are also special weatherproof switches which are designed for outdoor use.

Ceiling-mounted cord-operated switches are also available for use in bathrooms

# Raw Materials

or wherever the convenience of a cord-pull switch is required. They are rated at 5 or 6 amps.

Dimmer switches allow you to control the brightness of individual lights at the turn of a knob. One- and two-gang versions are available, usually the same size as a single plate-switch.

**Ceiling roses** These provide the connection between the lighting circuit and a pendant light. Modern roses have two or three sets of terminals, and the flex to the light can either be wired directly into it, or may be connected to a plug that fits into

a socket in the rose itself, allowing the light to be taken down easily for cleaning or repair. This latter type is known as a *luminaire support coupler*, or LSC for short. White plastic is the most common finish, but brass roses and LSCs are also widely available.

**Lamp-holders** As their name implies, lamp-holders hold lamps (the trade name for light bulbs). The pendant type is fitted to a pendant flex beneath a ceiling rose, while batten lamp-holders are designed for direct mounting to wall or ceiling surfaces. Batten lamp-holders fitted in

Fig 4 (*left*)
1 round decor lamp
2 small round ball lamp
3 single-cap tube lamp
4 mushroom GLS lamp
5 clear GLS lamp
6 pearl GLS lamp
7 coloured GLS lamp
8 long-life lamp
9 pygmy lamps
10 clear and pearl candle lamps
11 fireglow lamp for coal-effect electric fire.

Fig 5 (*below*)
1 internally-silvered (ISL) lamps
2 crown-silvered lamps
3 ISL lamps
4 parabolic aluminized reflector (PAR) lamps.

# Raw Materials

bathrooms must have a deep protective 'skirt' to prevent any risk of live parts being touched when changing a bulb.

**Junction and conduit boxes** These are components normally hidden above ceilings when the wiring is completed. They provide connections to individual light fittings and their switches.

## Light Bulbs and Tubes

There is now a huge range of light bulbs and tubes available for domestic lighting use – they come in many different shapes, sizes, colours and styles and divide into three broad categories.

**General service lamps** This group includes the standard 'round' and 'mushroom' bulbs most people use throughout the house, and also more exotic types such as candle lamps, pygmy bulbs and so-called 'decor' lamps which are intended for use with ornamental shades where the bulb may be visible. All come with clear or pearl glass, in a range of colours and wattages.

**Reflector lamps** These all have an internal silver coating which is designed to reflect the light beam in a particular way. Internally silvered lamps are silvered round the base and sides, so that the light is thrown forward in a broad beam. In crown-silvered lamps the top of the bulb

is silvered so the beam is thrown back into the fitting and then projected forwards by a reflector inside the fitting. Parabolic aluminized reflector (PAR) lamps have armoured glass, and are intended for outdoor use. Again, a range of colours and wattages is available.

**Tubes** Fluorescent and filament tubes come in several different sizes and wattages. Standard fluorescent tubes are 38mm in diameter, slimline tubes measure 25mm, while miniature tubes are generally 15mm across. Standard tube wattage range from around 20 watts for a 600mm-long tube up to 125 watts for a tube which is 2.4m long. Miniature tube wattages range from about 4 watts upwards. You can also buy circular tubes in two sizes – 300mm across rated at 32 watts, and 400mm across rated at 40 or 60 watts. Colours range from a bluish hue to a warm pinkish-white.

Miniature tungsten tubes are used mainly in decorative fittings such as picture lights, while tungsten halogen lamps in tube and bulb form are very popular for use in low-voltage light fittings run from a transformer.

You can now buy small upright fluorescent tubes, with or without a glass envelope, for use in ordinary light fittings. They offer longer bulb life and reduced electricity consumption.

**CHECK**
- the type of end cap on lamps and tubes when buying replacements. Most lamps have standard bayonet caps (BC for short) 22mm in diameter, with two contacts and two small locating pins, but some are also available with small bayonet cap fittings (SBC for short) 15mm in diameter

  The alternative for lamps is the Edison screw fitting. This comes in a wide range of sizes.

  Most fluorescent tubes have standard bi-pin end caps, but filament tubes usually have single-centre contacts.

Fig 6 (*left*)
1 SL lamp with fluorescent tube inside glass outer envelope
2 circular fluorescent tube
3 'architectural' tube with peg caps
4 double-cap filament tube
5 miniature fluorescent tube
6 slimline fluorescent tube
7 standard fluorescent tube
8 tungsten halogen tube
9 linear filament tube.

# Raw Materials

## Cable and Flex

Cable is used for all fixed wiring work, supplying the various circuits around the house. Flex is used for connecting pendant lights and appliances to the mains.

**Cable**  It is easy to recognize cable; it is fairly stiff, and has a flattened oval cross-section. The current-carrying conductors are protected within a tough PVC outer sheath, which may be grey or white (the latter is used for surface wiring).

Almost all your circuit wiring will be carried out using cable containing three cores. Two have colour-coded insulation – the red core is used to connect live terminals and the black core for neutral ones. The third core runs between the live and neutral cores, and is not insulated; this is used as the earth continuity conductor, and provides a continuous path throughout the system along which current can flow to earth in the event of an electrical fault. Where the core is exposed to connect the cable to any wiring accessory, this bare earth is always covered with a piece of green and yellow PVC sleeving.

You will need a special four-core cable if you intend to install any two-way switching arrangements (*see* pages 42 to 43 for more details). This has three insulated cores – coloured red, yellow and blue for indentification purposes – plus a bare earth core, and is used to link the two-way switches together.

Cable comes in several sizes, which are identified by the cross-sectional area of their conductors measured in square millimetres ($mm^2$). The most common and where they are used, are:

$1mm^2$ – for lighting circuits.
$1.5mm^2$ – for circuits rated up to 15 amps, such as to an immersion heater.
$2.5mm^2$ – for power circuits (except radial circuits rated at 30 amps – *see* page 59), and circuits to heaters rated up to 5kW.
$4mm^2$ – for circuits to small cookers and water heaters rated at up to 7kW, and for 30-amp radial circuits.
$6mm^2$ – for circuits to cookers, water heaters and instantaneous showers rated at over 7kW.

**Flex**  Flex is an abbreviation for flexible cord, which describes it exactly. As with cable, the cores are contained within an outer sheath, but there are several important differences with flex. Firstly, it comes with or without an earth core. Three-core flex is used for most applications, allowing the appliance concerned to be connected to the house earth continuity conductor for safety. Two-core flex with no earth core is used to wire up non-metallic light fittings, and also appliances such as hair driers, food mixers and power tools which are themselves double-insulated and so need no earth connection. Secondly, the cores are arranged differently – side by side in two-core flex, which is available with both flat and round sheathing, but arranged in a triangle in three-core flex, which is always round in cross-section. Thirdly, the earth core is itself insulated.

The cores are colour-coded brown for live, blue for neutral and green and yellow for earth. You may find that flex fitted to old appliances has the old colour codes – red for live, black for neutral and green for earth.

Most flex comes with PVC sheathing, generally white but also available in black and orange (and other colours to special order). You can also buy so-called un-kinkable flex, which has a vulcanized rubber sheath covered with braided fabric and is popular on appliances such as irons or heaters. Heat-resisting flex has a butyl rubber sheath and is used for wiring immersion heaters and other water heaters, and some enclosed light fittings.

Special multi-core flex is used for wiring up central heating controls, and parallel twin flex is used for door bells and loudspeaker connections.

Flex, like cable, is identified by the cross-section of the conductors. The most common sizes and their uses are:

$0.5mm^2$ – used for wiring light fittings.
$0.75mm^2$ – for light fittings and small appliances rated at up to 1.4kW.
$1mm^2$ – for appliances up to 2.4kW.
$1.5mm^2$ – for appliances up to 3.6kW.
$2.5mm^2$ – for appliances up to 4.8kW.
$4mm^2$ – for appliances over 4.8kW.

**Note**  For more information on cable and flex, *see* pages 21 to 27 and 76 to 79.

**WIRING SUNDRIES**
There are several odds and ends you will need for wiring work. The list includes:

- green and yellow PVC earth sleeving, for protecting bare earth cores within wiring accessories
- red PVC insulating tape, for general insulation work and for identifying live cores on switch circuits
- rubber grommets, for fitting into knock-outs in flush metal mounting boxes to prevent cable sheaths from chafing on bare metal.
- strip connectors, for making connections within conduit and junction boxes
- cable clips

**CHECK**
- that flex used for pendant lights is strong enough to support heavy lampshades. The weight each size can carry is:

  $0.5mm^2$   2kg (4½lb)
  $0.75mm^2$  3kg (6½lb)
  $1mm^2$     5kg (11lb)

**CABLE CORES**
From 31 March 2006, you **must** use cable with the new core colours for all wiring work. This cable is already available and can be used now. *See* page 6 for full details.

# Tools and Equipment

You need a small and comparatively inexpensive selection of specialist tools for carrying out wiring work. In addition, you will find a use for many of your general-purpose carpentry and building tools, and it is well worth keeping a small emergency tool-kit somewhere handy to help you cope with the unexpected.

## Specialist Tools

The tools you need are mainly for cutting and stripping flex and cable, for making connections to wiring accessories and for routing cables in walls and under floors around the house.

**Side-cutters** These are essential for cutting cable and flex to length, although at a pinch you could use the cutting jaws in a pair of combination pliers. You can also use the cutters for trimming cable and flex cores to length when making connections to terminals. Choose a pair with insulated handles, for comfort as well as safety.

**Wire strippers** These are equally essential, and are used for stripping the insulation from cable and flex cores without damaging the cores themselves. This is especially important with flex, where cutting through the fine conductor strands can seriously reduce the current capacity of the flex, leading to overheating. The strippers are adjustable, allowing you to set the jaw separation to match the conductor size exactly. Again, make sure they have insulated handles.

**Pliers** These are used to grip and bend wire, and to twist pairs of cable cores together before they are connected to accessory terminals. Special electrician's pliers have insulated handles. It is also useful to have a pair of long-nosed pliers, for tucking cores into accessory terminals and so on.

**Screwdrivers** These are needed in various sizes for everything from tightening terminal screws to fixing accessories to walls and ceilings. Buy a thin-bladed electrician's screwdriver for terminal screws, ideally with an insulated blade. You will need a medium-sized screwdriver for driving face-plate fixing screws (which are always of the slotted-head variety), and a large screwdriver for fixing mounting boxes and the like.

It is worth using cross-headed screws (driven by a matching screwdriver) for

Fig 7 (*below*)
1 torch
2 flooring saw
3 floorboard chisel
4 wire strippers
5 side-cutters
6 long-nosed pliers
7 electrician's pliers
8 joist brace
9 continuity/test meter
10 neon tester
11 plug-in socket tester
12 electrician's screwdrivers.

fixing mounting boxes and the like, as with these you can balance the screw on the tip of the blade and offer it up to the fixing within confined spaces. You may find it worth investing in a ratchet screwdriver, or even a cordless power screwdriver if you are planning extensive wiring work.

**Test equipment** Professional electricians use a range of test equipment which is not readily available to the do-it-yourselfer, and in any case certain tests are best left to the professional (*see* page 86). However, there are three useful items of test equipment you should have in your electrical tool-kit.

The first is a *neon tester*, which resembles an electrician's screwdriver but has a small neon indicator built into the handle to indicate if a component is live. If you touch the insulated blade of the tester to the component and press the cap on top of the handle, a circuit is completed which lights the tester bulb if the component is live. A built-in resistor prevents

you from receiving any shock when testing live components.

The second is a *continuity tester*, which is used to check that there is a continuous current path in parts that cannot be inspected visually – for example, to test whether a flex core is undamaged, or whether a cartridge fuse has blown. Some come in the form of test meters, others simply have a test light that indicates if there is circuit continuity when the probes are touched to the flex or fuse ends.

The third is a *socket tester*, a special plug designed to be inserted in 13-amp socket outlets to check whether the socket is wired correctly, or has faulty or reversed connections.

**Flooring tools** These are used for running cables beneath timber floors. You will find it useful to have a flooring saw for cutting through the tongues along board edges, and a floorboard chisel for lifting them. A joist brace makes it easy to drill holes through joists in confined spaces. Use power tools if you prefer.

Fig 8 (*below*) You may need many of your general-purpose tools for wiring work, plus plentiful supplies of fixing devices – screws, wall-plugs and cavity fixings of various types.

# Tools and Equipment

## General-Purpose Tools

You will need many of your general building and woodworking tools for wiring work.

An electric drill plus a range of twist and masonry bits is a must for a variety of drilling tasks. Have a hand drill available too, for occasions when you have to drill holes and the power is off.

A club hammer plus a brick bolster and several different sizes of cold chisel will be needed for cutting out recesses for mounting boxes and for cutting chases in masonry for cable runs. If you have a lot of work to do, consider hiring a power chasing tool; it cuts chases quickly and effortlessly, but it does make a lot of dust!

Chisels and saws are useful for cutting notches in joists and stud wall timbers. It's worth having a pad-saw for cutting holes in plasterboard or lath-and-plaster ceiling and walls, and a junior hack-saw comes in handy for sundry jobs like cutting conduit.

## An Emergency Tool-Kit

However well organized you are, it is always difficult to find the tools and equipment you need in an emergency – especially when the house is plunged into darkness by an electrical fault or malfunction. Therefore it is well worth assembling a simple selection of tools and sundries, packed in a small container and kept somewhere accessible inside the house so you can lay hands on it quickly when you need it.

Your emergency tool-kit should contain a couple of screwdrivers (including one small electrician's screwdriver), a pair of pliers, a handyman's knife, a torch (plus a set of spare batteries for it), a roll of PVC insulating tape, and either a card of fuse wire or a set of replacement cartridge fuses, depending on what sort of circuit fuses your system has. Make sure you also have some spare plug fuses in 3-amp and 13-amp ratings. Finally, if space permits, add two or three spare light bulbs.

Note   For more information on electrical repairs, *see* pages 72 to 82.

Fig 9 (*left*)  Pack a suitable container with basic tools and wiring accessories, and keep it somewhere handy for coping with emergency electrical repairs.

# Shopping for Electrical Goods

Where best to shop for electrical fittings and equipment depends largely on the scale of your operations. If all you want is a plug and some fuse wire, look no further than your local hardware store, but if you are working on a larger scale you will make considerable savings on your final bill by shopping around. Here are the places to try.

## Local DIY Shops

The typical independent high street DIY shop usually stocks a small selection of electrical goods. You'll be offered a choice of just one brand of wiring accessory, probably blister-packed, and in a range extending only to obvious things like switches, ceiling roses and socket outlets. It will also stock the more common types of cable and flex, which will be sold by the metre and rather expensively. It will probably not have all the specialist tools you need.

**Verdict** Fine for small jobs such as wiring a spur socket, where all you need is one accessory and a metre of cable, and do not mind paying more than you need to for the convenience of shopping locally; expensive for larger projects.

## DIY Superstores

The major national chains generally offer a good range of wiring accessories, tools, materials and sundries, although you may be restricted in choice of brand. Some offer parallel ranges of branded and own-brand accessories, often blister-packed and including things like small consumer units with MCDs. Cable and flex is sold by the metre or in complete reels. One or two chains also offer useful advice leaflets on simple electrical jobs.

**Verdict** A good selction of tools and materials at fairly reasonable prices, but unlikely to offer much technical advice.

## Specialist Retailers

There is a growing number of specialist electrical retailers around, offering an excellent range of accessories and other electrical equipment, and many are also happy to offer sound and useful technical advice. You should be able to buy everything you need for even the most complex wiring work, plus any specialist electrical tools. Lighting retailers offer by far the biggest choice of light fittings

**Verdict** Excellent selection of equipment and specialist tools, generally at reasonable prices; good for technical advice too.

## Builders' Merchants

Most usually stock a comprehensive range of fittings, accessories, sundries and tools, but may be loyal to only one brand of accessory. Some have 'retail' counters to cater for the non-trade customer.

**Verdict** Reasonable selection of goods at a reasonable price.

## Electrical Wholesalers

Specialist wholesalers stock by far the widest range of wiring accessories and other electrical equipment, often offering several different brands. However, since they exist to cater for the professional electrician, you need to know what you want before you start. It is worth contacting local firms for estimates if you plan any large-scale wiring work.

**Verdict** By far the biggest choice of materials and tools and generally at the keenest prices; good for large orders.

## Mail Order and Online Suppliers

You can now obtain all the electrical tools and equipment you need from a growing number of mail order and online suppliers such as Screwfix (www.screwfix.com).

**Verdict** Prices are extremely competitive and delivery generally very prompt.

---

**CHECK**
- that electrical goods you buy conform to the relevant British Standards; the BS number should be stamped on the rear of the accessory face-plate. Below are the standards to lok out for:

- plugs BS1363
- plug fuses BS1362
- socket outlets BS1363
- shaver sockets BS4573
- FCUs BS1363
- DP switches BS3676
- plate switches BS3676
- dimmers BS5518
- junction boxes BS6220
- ceiling roses BS67
- plug-in roses BS7001
- lamp-holders BS EN60238/61184
- metal mounting boxes BS4662
- plastic mounting boxes BS5733
- consumer units BS5486 or BS EN60439
- circuit fuses BS1361
- MCBs BS3871 or BS EN60898
- RCDs BS4293 or BS EN61008

**TIP**
Buy in bulk for fittings and materials you will use a lot, especially circuit cable (sold in 50m and 100m drums) and accessories such as double sockets (sold in bulk packs of ten).

# Electrical Safety

This is the most important page in this book. Read it every time you plan to carry out any electrical work.

## Safety First

It goes without saying that electricity can kill, and you are not only at risk when you carry out alterations or repairs to your wiring system. There are also everyday risks to you and every member of your family if something goes wrong with your system or if you fail to do something properly in the first place. Below are ten golden rules to obey at all times:

1.  Do not attempt any electrical work unless you know exactly how to carry it out and are confident of your technical ability to do so.
2.  Always turn off the power at the main switch before you start any electrical work on any part of your system.
3.  If you are working on just one circuit, remove the appropriate circuit fuse-holder and keep it with you until you have finished so no one can replace it without your knowledge. If you have MCBs, put tape over the switch after turning it off. Never try to mend a blown circuit fuse with a metallic object such as a nail or silver foil; you may succeed in restoring the current flow, but the circuit will have no fuse protection and someone could be killed if a fault develops.
4.  Always double-check all connections to make sure that cores go to the right terminals, that they are securely anchored and that no bare conductors are visible.
5.  Never omit the earth connection if one is needed. With flush metal mounting boxes, always fit an extra earth core between the accessory face-plate and the earth terminal in the mounting box.
6.  Never touch any appliance or fitting with wet hands, or use electrical equipment in wet conditions. In particular, never take any portable appliance into the bathroom using a mains extension lead.
7.  Always unplug an electrical appliance before attempting to carry out any repairs on it. If an appliance is wired to a fused connection unit or double-pole switch, turn off the isolating switch first.
8.  Don't use long trailing flexes – they can be trip hazards. If you have to extend a flex, use a proper flex connector. Always uncoil extension leads fully from their drums to prevent the flex overheating. Avoid overloading socket outlets with adaptors; if you need to use adaptors, you need more outlets.
9.  Check the plug connections and the condition of the flex on all portable appliances regularly (at least every six months), remaking loose connections and replacing damaged flex if necessary. Replace damaged wiring accessories or plugs as soon as possible.
10. Teach your children about the dangers of electricity as soon as they are old enough to understand. Until they are, fit all socket outlets with special socket covers, and avoid leaving appliances with trailing flexes plugged into the mains.

## Safety Devices

The best safety device you can fit to your system is a residual current device (RCD), which will protect both your system and its users. The biggest danger is if someone touches a live component somewhere on the system and receives an electric shock; if severe, it could be fatal. An RCD will shut off the power supply within milliseconds if it detects the fault resulting from current flowing to earth through your body – fast enough to prevent the shock from stopping your heart beating. It will also detect any current leakage to earth caused by faulty insulation – the sort of fault that can start an unseen electrical fire.

Fitting a whole-house RCD is a job for a professional electrician. If your system does not have one, get one without delay.

For extra protection, always use a high-sensitivity RCD socket outlet or plug-in RCD adaptor to power tools or electrical appliances when carrying out jobs out of doors.

---

**FIRST AID**

If you or your family member receives a shock, however slight, from an electrical appliance, stop using it at once. If the shock is from a switch or socket outlet, check the circuit wiring yourself or call in an electrician.

In cases of severe shock, try to turn off the supply. Don't touch the person if he or she is still gripping the source of the current; use a non-conductor such as a wooden broom to separate the victim from the source.

Lay the victim flat if conscious but visibly shocked, cover burns with a dry sterile dressing and call a doctor. Do not give food, drinks or cigarettes.

Lay an unconscious casualty in the recovery position, keep the airway clear and *call an ambulance immediately*. Check breathing and heartbeat, and give artificial respiration if either of these stops.

# Stripping Flex and Cable

The key raw materials for all your wiring work are flex and cable. Remember that flex is used for connecting portable appliances to the mains via plugs and sockets, and also for linking pendant lamp-holders to their ceiling roses. Cable is used for all circuit wiring work, and also uniquely for connecting free-standing cookers to the mains; flex is used for all other appliances. Before they can be used to make these connections, you have to strip back the outer sheath and the core insulation to reveal the bare conductors inside.

The sheath is there to protect the cores from damage. You should therefore aim to remove just enough of it to allow the cores to reach the terminals inside the plug or wiring accessory to which they are being connected, while ensuring that the sheath runs unbroken into the body of the plug or accessory. Similarly, the core insulation should run right up to each terminal, with just enough bare conductor exposed to allow a secure connection to be made.

## What to do

**PVC-sheathed flex**   Mark the sheath at the point where you want to sever it. Slit it carefully lengthwise (Fig 10), then peel it back and cut it off. If you have nicked the core insulation, cut off the cores and try again. Next, cut the cores to length so each will reach its terminal. Then use wire strippers (Fig 11) to remove about 12mm (½in) of the core insulation taking care not to cut through the conductor strands. Twist the strands of each core together neatly.

**Fabric-covered non-kink flex**   Slit the fabric cover first and cut it away. Then slit the rubber sheath and remove this too. Prepare the cores as before.

**Cable**   Either slit the sheath lengthwise with a knife as for flex (Fig 12), or grip the bare earth core with pliers and pull it backwards to slit the sheath. Cut away the sheath, strip the core insulation with wire strippers and fit PVC sleeving to the bare earth core (Fig 13).

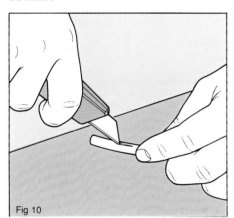

Fig 10

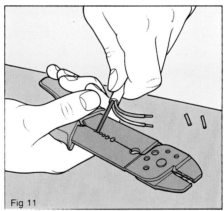

Fig 11

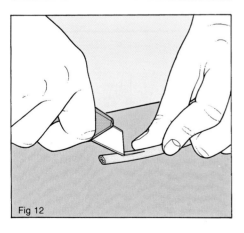

Fig 12

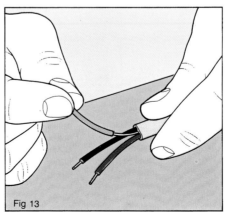

Fig 13

Fig 10   Use a sharp knife to slit the flex sheath.

Fig 11   Cut off the excess and prepare the cores with wire strippers.

Fig 12   Cut cable sheath along the centre line.

Fig 13   Remove the excess, prepare the cores and add earth sleeving.

21

# Wiring a Plug

The vast majority of homes now have socket outlets that accept the familiar 13-amp fused plug with its three rectangular pins. Unfortunately, few appliance manufacturers supply their products fitted with a plug (unlike their Continental competitors), so for the foreseeable future you will be faced with the prospect of fitting one yourself. Since the plug is the vital link between appliance and power supply, it is important to know how to wire one up correctly.

All plugs now on sale must have sheathed live and neutral pins, to reduce the risk of slim fingers touching live metal as the plug is withdrawn from its socket. They also have markings to indicate which terminal is which, but to help you remember there's a simple mnemonic. With the plug open and facing you, always connect the BRown (live) flex core to the Bottom Right terminal, and the BLue (neutral) core to the Bottom Left one. The earth core (if you are using three-core flex) goes to the top terminal.

## What to do

Start by preparing the flex (see page 21), removing about 38mm (1½in) of the sheath. Then open the plug and lay the flex over it with the sheath positioned on top of the cord grip. Take each core along the channel towards its terminal so you can see how long it needs to be, and cut it to length if necessary. Then use your wire strippers to remove about 12mm (½in) of the insulation from each core.

With **pillar terminals** (Figs 14, 17 and 18), loosen the terminal screw, insert the bare conductor in the hole and tighten the screw to secure it. With **stud terminals** (Figs 15 and 19), remove the nut, wind the conductor clockwise round the stud and tighten the nut and washer down. With **snap-down terminals** (Fig 16), simply lift the clamp, lay the cores in place and press the clamp down.

Secure the sheath in the cord grip, check that the right fuse is fitted (see Check) and screw the plug top on.

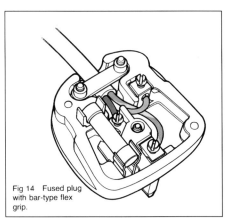

Fig 14   Fused plug with bar-type flex grip.

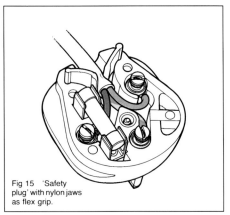

Fig 15   'Safety plug' with nylon jaws as flex grip.

Fig 18   (*above*)   Pillar-type terminal.

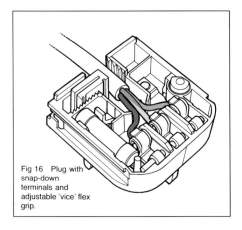

Fig 16   Plug with snap-down terminals and adjustable 'vice' flex grip.

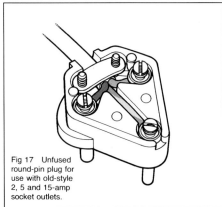

Fig 17   Unfused round-pin plug for use with old-style 2, 5 and 15-amp socket outlets.

Fig 19   (*above*)   Stud-type terminal.

# Planning Cable Runs

If you want to alter or extend your home's wiring in any way, you will have to decide how you are going to run the cable from A to B – for example, to supply an extra socket outlet wired up as a branch from an existing power circuit. The shortest route is obviously the best in terms of economy of cable, but practical considerations have to be taken into account too when you are selecting the best route – you do not want to demolish a room just for the sake of hiding a new cable.

When a new house is wired up, all the cables are concealed behind wall surfaces or under floors before plaster is applied or floorboards are fixed down. Ideally you should also conceal any new wiring you do, since this means it is out of sight and out of harm's way. However, you may not want to ruin your decorations or lift fitted carpets, so some sort of surface mounting is often a good short-term solution; you can always re-route and conceal the cable in the future, when you next strip the room for redecoration.

## What to do

If you plan to extend your wiring, your first step is to decide what you want to add and where it is going to be sited. You then need to examine the structure of your house so you can work out the cable route.

Cable runs to wall-mounted accessories such as wall lights, switches and socket outlets are usually taken vertically up or down the wall in channels (called chases) cut in the plaster. If necessary, they can also run horizontally, but should *never* be run diagonally – once concealed, no one would be able to deduce where they ran, and they could be pierced when making wall fixings. In stud partition walls, cables are run in the gaps between the vertical frame timbers.

In houses with timber floors and plasterboard ceilings, cables are run between and across the joists, and can be run in any direction. With solid floors, it is better to route new cables round the perimeter than to cut channels in the floor surface.

**CHECK**
- whether walls are built of solid masonry, or are timber-framed partitions clad with plasterboard. Beware of walls in old houses, which may have a lath-and-plaster surface over solid masonry
- which way floorboards run. Their supporting joists will run at right angles to the boards, so you can easily work out whether your proposed new cable run will be parallel to or across the joist line
- whether you have easy access to underfloor and ceiling voids. Built-in furniture or newly added partition walls may make it difficult to lift certain floorboards

**TIP**
Use a small battery-powered metal detector to check for existing wiring or plumbing pipework buried in wall and solid floor surfaces. It will also detect the lines of nails holding plasterboard to ceilings and stud partition walls, so helping you to locate the positions of joists and frame members.

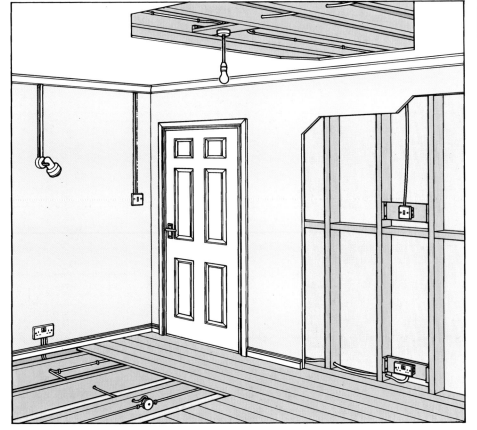

Fig 20 Your house may have concealed or surface-mounted wiring. Some typical cable routes are shown here.

# Mounting Cable on Surfaces

Fixing circuit cables to wall and ceiling surfaces is certainly the quickest way to extend your wiring, and many homes have some of their wiring carried out in this way. However, surface-mounted wiring has two main drawbacks. It can look unsightly, and the cables are also more prone to accidental damage than if they are concealed within the structure.

One way round this problem is to run the cables in protective plastic channelling (often called mini-trunking), which is itself fixed to the wall or ceiling surface. Systems of this sort have snap-on cover strips, and offer a full range of straight connectors, tees, elbows and so on to enable you to assemble complete runs with ease. White is the most common colour, but other colours are available.

An alternative is to use larger types of trunking which replace conventional skirtings, architraves and cornices, allowing you to run cables round rooms at floor and ceiling level and round door openings.

## What to do

If you want to run bare cables to new lights or wiring accessories, perhaps as a temporary measure until you redecorate the room and can then conceal the cable runs, simply secure them in place with cable clips at about 300mm intervals. Clips are sized to match common cable sizes, and most are fitted with masonry pins for use on solid walls. Where possible, run the cable along the tops of skirting boards and up the architraves round doors.

With mini-trunking, simply stick, pin or screw the U-shaped base sections to the wall or ceiling surface, adding fittings as necessary when the run changes direction. Then lay in the cables and snap on the cover strip to retain them. Cornice trunking is fitted in the same way.

With skirting and architrave trunking, you generally have to remove the existing mouldings first. Fix the new system as for mini-trunking.

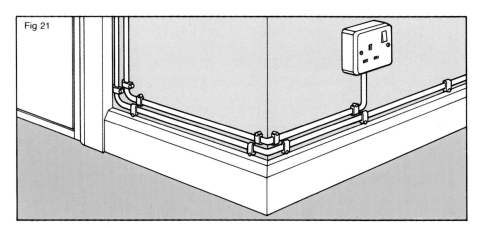

Fig 21 It is quick and easy to mount cable on a surface with clips, but the result is unsightly and the cables are prone to accidental damage.

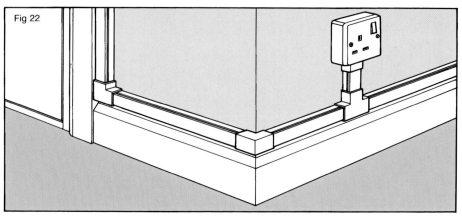

Fig 22 Surface-mounted mini-trunking is neater and also protects the cables from damage. Trunking is available in a wide range of sizes and styles.

> **TIP**
> If you are planning to use skirting or architrave trunking, fit a system with spare capacity so you can extend the wiring easily in the future.

# Running Cable across Floors

If you have suspended timber floors, one of the easiest ways of running cables from one side of the room to the other is to use the void underneath the floorboards. Where the cable run is parallel to the joist direction and there is a ceiling below, it's generally possible to lift a floorboard at each side of the room and feed the cable through. Beneath suspended timber ground floors, the cable should be clipped to the joists – from below if there is sufficient crawl space, or by lifting intermediate boards otherwise. Where the cable crosses the joist line, it can run through holes drilled in the centres of the joists, or in shallow notches cut in their top edges and protected from damage by metal repair plates – the former method is better.

## What to do

Start by lifting floor coverings to expose the floorboards. Then prise up floorboards as required (you may need to use a floorboard saw or circular saw to cut through the tongues of interlocking boards). With runs parallel to joists, simply push the cable through between them (but *see* Check). For runs across the joist line, lift one row of boards and use an electric drill or a brace and bit to make holes in the joist centres so you can pass the cable through.

If you have chipboard floors, start by cutting through the tongues along the joint lines. Then punch the fixing nails through the boards so you can prise them up without snapping them.

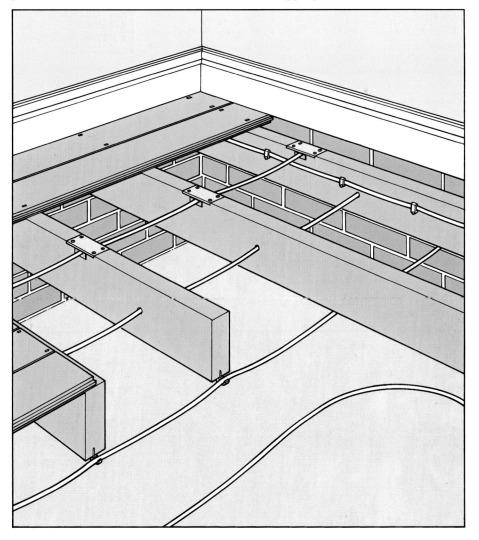

Fig 23  Beneath timber floors, cable can run across the joists or parallel to them as shown. Unless notches and protective plates are used, the cable should always be at least 50mm (2in) below the top edge of the joists.

# Running Cable in Solid Walls

Unless you live in a timber-framed house, most of your walls – downstairs, at least – will be of solid masonry, and the neatest way of running cables across them is to cut channels called chases in the wall surface. The cable is then secured in the chase with cable clips and is plastered over. If you want to give the cable extra protection, or if you foresee wanting to alter or extend the wiring using the same cable route, you can enclose it in flat PVC conduit which is itself embedded in the chase. However, this is not an essential requirement.

In older homes, the plaster may be very thick, enabling you to cut the chase without having to cut into the brickwork beneath. However, modern homes have much thinner plaster so if you want to use PVC conduit you may be faced with the harder task of chiselling out brickwork or blockwork also (see Tip). Remember that cable runs in walls must travel vertically or horizontally, but must never run diagonally.

## What to do

Once you have decided where the cable will run, start by marking the position of the wiring accessory it will be supplying, and then mark the route of the cable run to the accessory (Fig 24). Use a sharp cold chisel and a club hammer to cut along each side of the chase and to remove the plaster between the cut lines (Fig 25). Where the chase crosses picture rails and dados (chair rails), use a masonry bit in your electric drill to clear out the plaster behind (Fig 26). You can use the same technique to clear a chase behind a skirting board if you want the cable to pass up the wall from floor level.

If you plan to use conduit to protect the cables, cut it to length and check that it fits snugly within the chase. It should stop about 12mm (½in) above the position of the wiring accessory. Secure it in place with galvanized plasterboard nails, ready for the cable. Finally, chop out the recess for the accessory (see page 29).

see page 29

## What you need:
- plumb-line
- straight edge (ruler)
- pencil
- club hammer
- sharp cold chisel
- electric drill plus masonry bit
- flat PVC conduit
- junior hack-saw
- galvanized nails
- claw hammer
- work gloves

## CHECK
- that no old wiring or pipework is hidden beneath the plaster by using a battery-powered metal detector
- that old plaster is sound, by tapping the surface. If it sounds hollow, beware: cutting the chase could bring large sections of plaster off the wall. Look for another route, or consider surface-mounting the cable instead

## TIP
If you have a lot of chase cutting to do, consider hiring a special power tool which will cut neat straight-sided chases for you with minimal effort.

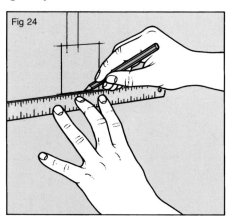

Fig 24

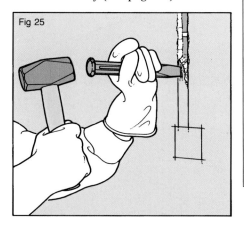

Fig 25

Fig 26

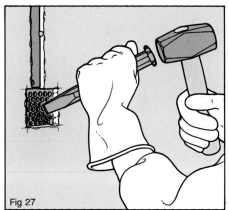

Fig 27

Fig 24   Mark the cable run and accessory position.

Fig 25   Chop out the chase with hammer and chisel.

Fig 26   Drill out plaster behind picture rails.

Fig 27   Fit conduit and chop out the accessory recess.

# Running Cable in Hollow Walls

Many homes have internal walls consisting of sturdy timber frames covered on each side with plasterboard (or with lath and plaster in older homes), and modern timber-framed houses also have this type of construction for the inner leaf of their cavity walls.

This type of hollow wall provides the ideal route for cables, although once the wall has been constructed it can be tricky to get the cable to where you want it without having to cut away some of the plasterboard first. In timber-framed houses, the exterior walls contain a vapour barrier, so new cables cannot be concealed within them without damaging it.

## What to do

Start by locating the positions of the wall studs (verticals) and noggins (horizontals) – either by tapping the wall, or by using a metal detector to locate the fixing nails. Then cut away panels of plasterboard with a sharp knife where new accessories are to be fitted, and where the cable run will cross a nogging or stud. Screw supporting battens between the studs to provide fixings for the accessories (see page 30 for more details), and drill holes or cut notches where necessary to allow the cables through. When wiring is complete, nail the plaster-board panels back.

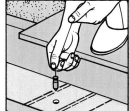

Fig 29 (*above*)  Locate the head plate of the partition wall, drill a hole in it and lower a plumb-line.

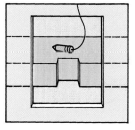

Fig 30 (*above*)  Where the bob hits a nogging, cut away the plasterboard and notch the timber. Then tie the cable to the plumb-line and use this to draw the cable into place.

Fig 28 (*left*)  Stud partition walls are ideal for concealing cables, run from below or above.

# Mounting Accessories on Surfaces

Whether your circuit cables are concealed or surface-mounted, the easiest way to fit new wiring accessories is to mount them on the wall surface. Every type of accessory must be mounted over an enclosed box so that the wiring is inaccessible. With surface-mounted accessories, the extra depth of the plastic or metal mounting box screwed to the wall surface means that the fitting projects noticeably from the wall, especially with power circuit accessories such as socket outlets. This is a potential drawback, but one which you may be prepared to put up with as an alternative to chopping holes in your walls for recessed metal mounting boxes (see page 29 for more details).

You may also prefer to fit surface-mounted accessories if you are doing wiring work and do not want to damage the wall decorations at this stage; you can then recess the fittings (and any surface-run cables) the next time you redecorate the room.

## What to do

Start by deciding where you want the new accessory to be fitted, and hold its plastic mounting box against the wall so you can mark the positions of the fixing screws (Fig 31). Then decide how the cable will be run to the accessory, and punch out one of the thin plastic sections (called knock-outs) from the back of the box to allow the cable to enter (Fig 32).

On solid walls, drill holes (Fig 33) and fit wall-plugs so you can screw the box to the wall. On stud partition walls, you must screw the box to a vertical stud; cavity wall fixings are not strong enough, and will pull out of the plasterboard as plugs are withdrawn from sockets.

With the box in place, run cable to it – either recessed in a chase or surface-mounted – and feed it into the mounting box through the knock-out (Fig 34), ready for preparation and connection to the new wiring accessory.

**What you need:**
- pencil
- small spirit level
- screwdriver
- electric drill plus masonry bit
- wall-plugs
- wood screws
- cable
- cable slips *or* surface trunking *or* flat PVC conduit
- side-cutters
- mounting box

**CHECK**
- that the box is level using a small spirit level. The fixing holes are elongated to allow minor adjustments
- that on stud partition walls you screw the box to one of the timber uprights, not into the plasterboard
- that the mounting box is deep enough to accept whatever accessory you want to fit to it. You need boxes 17mm deep for light switches, 32mm deep for most power circuit accessories and 44mm deep for some special controls

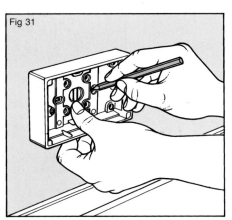

Fig 31

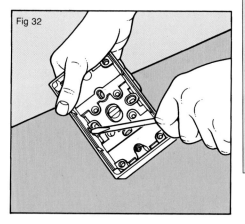

Fig 32

Fig 33

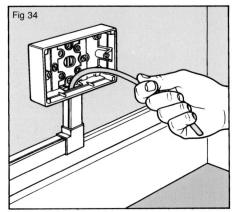

Fig 34

Fig 31   Mark the position of the box fixing holes on the wall surface.

Fig 32   Use a screwdriver to remove a knock-out from the base of the box.

Fig 33   On solid walls, drill and plug the fixing holes and fit the box.

Fig 34   Screw the box in place and feed the cable in through the knock-out.

# Mounting Accessories in Solid Walls

Undeniably the best way to mount wiring accessories is to fit them over recessed mounting boxes, so that only the accessory face-plate projects from the wall surface. The cable run to recessed boxes should also be recessed into the wall surface. This not only looks neat; it also helps to protect the face-plate from accidental damage. However, it also means more work for you, since you have to chop out some of the masonry behind the plaster – fairly easy if your walls are built in soft insulating blockwork, but hard work if they are built with dense bricks.

It is important to remember that the masonry in internal walls and the inner leaf of external cavity walls is generally only 102mm thick, so if you are too heavy-handed with your club hammer you may chop a hole straight through it, and have some awkward repair work to carry out as a consequence. The technique shown here is designed to minimize any risk of that occurring.

## What to do

As before, start by marking the position of the new accessory on the wall surface, and also the line of the cable run to it. Then use a masonry bit in your electric drill to honeycomb the masonry with a series of closely-spaced holes, drilled to a depth just greater than that of the box (Fig 35).

Next, use a sharp cold chisel and your club hammer to chop out the honeycombed waste to the required depth (Fig 36). Chop out the cable chase too, and test the box for fit in the recess. Chisel away the base a bit more if necessary to get a good fit.

Push out one of the circular knock-out plates in the side or back of the box (Fig 37) to allow the cable to enter, and fit a grommet (rubber washer) in the hole to stop the cable chafing on the metal.

Mark the positions of the box fixing holes, drill and plug the holes and screw the box in place (Fig 38). Run in the cable and finish off by filling around the box.

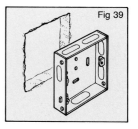

Fig 39

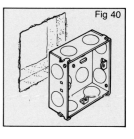

Fig 40

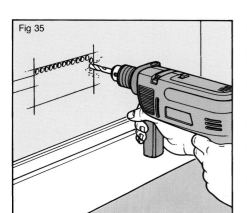

Fig 35

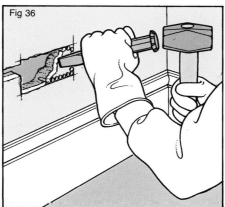

Fig 36

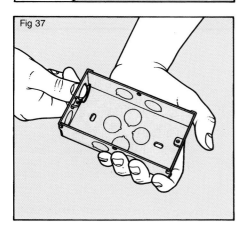

Fig 37

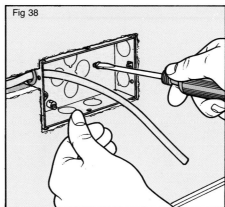

Fig 38

Fig 35   Mark the cable run and box position, and drill out the masonry.

Fig 36   Chop out the waste with a chisel and hammer.

Fig 37   Remove a knock-out and fit a grommet.

Fig 38   Drill and plug the wall, screw in the box and feed in the cable.

# Mounting Accessories in Hollow Walls

If your home has interior partition walls – timber frames clad with plasterboard or, in older homes, lath and plaster – you will need to adopt different techniques to those used on solid masonry walls if you want to mount recessed wiring accessories.

The most important thing is to achieve a firm fixing, since an accessory that pulls away from the wall could leave live wiring dangerously exposed. There are two traditional methods of doing this. Both involve cutting a 'window' in the wall surface, then either a notch is cut in a stud or nogging to accept the recessed box, or else a batten is mounted between two adjacent studs to which the box is then screwed. In each case, the plasterboard cut-out is nailed back and the joints made good.

A new, neater solution is to cut a smaller opening and to fit a special flush mounting box with flanges at each side that grip the rear face of the plasterboard to make a secure fixing.

## What to do

To mount a box on a batten, first locate two adjacent studs. Use a sharp knife to cut through the plasterboard along the centre lines of the studs, and a pad-saw to cut across the board between them. Lift out the panel and set it aside for re-use.

Now screw support blocks to the sides of the studs and secure a batten to them, set far enough back from the face of the wall to leave the lip of the box flush with it. Screw the box to the batten, remove a knock-out and feed in the cable (Fig 41).

To mount a box on a stud, cut out a window over it to match the box size. Then chisel away a recess in the stud deep enough to take the box, and screw it in place. Feed in the cable as before (Fig 42).

With flanged boxes you make a cut-out to match the box size anywhere on the wall except over a stud. Feed in the cable, press the box into the cut-out and check that the flanges are properly located (Fig 43).

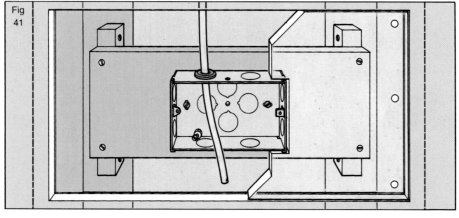

Fig 41

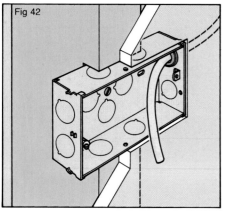

Fig 42

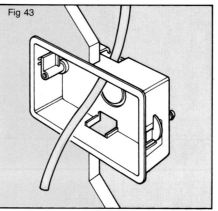

Fig 43

Fig 41   Cut a window in the wall and fit a batten to support the box.

Fig 42   Alternatively, fit the box in a recess cut in a stud.

Fig 43   Use flanged boxes to fit accessories anywhere.

# Fixing Accessories on Ceilings

Most homes have ceilings formed by nailing plasterboard to the underside of the ceiling joists; in older homes, lath and plaster may have been used instead. Neither is strong enough to support things like light fittings and ceiling switches. Instead, fixings must be made either directly to the underside of a joist, or else to a support batten mounted between adjacent joists – a technique very similar to that for fitting accessories in stud partition walls.

Ceiling roses supplying ordinary pendant lights are designed so that the knockouts are offset, allowing the rose to be fixed to the joist with room for the cables to enter it (Fig 44). Some types of light fitting such as fluorescent tubes, to which the circuit cables are directly connected, can also be screwed to a joist if one coincides with the desired light position. However, for many types of light fitting you have to provide a recessed enclosure within which the wiring connections are made.

## What to do

For roses and fittings screwed directly to a joist, all you need to do is to ensure that you locate the centre line of the joist, so that you can be certain that the screws go into solid timber. Then simply feed the circuit cable in from above (Fig 44).

For many other fittings, or where a rose must be fitted between the joists, you must fit a supporting batten. This will necessitate lifting floorboards in the room above for access. Where the fitting requires a recessed enclosure (see also page 37), you have to install a conduit box so its lip is flush with the ceiling surface.

At the desired light position, draw round the outline of the box (Fig 45). Drill a hole in the ceiling and insert a pad-saw blade so you can cut round it (Fig 46). Gain access from above, fit a batten on blocks between the joists (Fig 47), and screw the conduit box to the batten from below. Finally, feed the cable into the box.

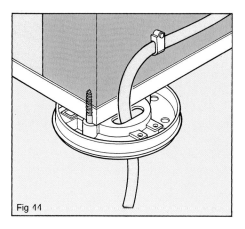

Fig 44

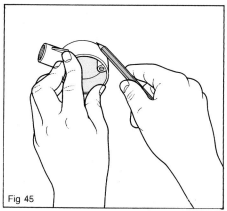

Fig 45

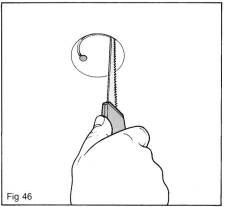

Fig 46

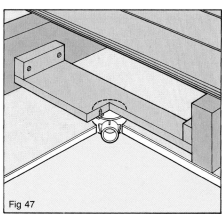

Fig 47

Fig 44   Where possible, screw ceiling roses directly to a joist.

Fig 45   To recess a conduit box into the ceiling, first draw round it.

Fig 46   Then drill a pilot hole and cut along the line with a pad-saw.

Fig 47   Gain access from above and fit a batten on support blocks to the joist sides. Then attach the conduit box to it.

# Connecting Cable to Accessories

Ensuring good connections between the circuit cables and the wiring accessories they are supplying is the most important part of any electrical job. Bad connections can lead to short circuits, overheating and ultimately electrical fires, and could also cause you or another member of your family to receive a shock from the accessory concerned.

The first golden rule is to be generous with cable when running it between two points; it's easy to cut off any excess, but impossible to put a bit back. Furthermore, a little slack in the cable actually makes it easier to connect the cable cores to the terminals within the accessory.

The second rule is to trim only enough core insulation to allow the core to make a good connection within its terminal. Bare wires cause short circuits if they touch, and shocks if they are touched.

The third rule is *always* to cover the cable's bare earth core with a length of green and yellow PVC sleeving, long enough to run right up to the terminal (Fig 48).

## What to do

Start by feeding the cable into its mounting box, and use side-cutters to cut off any excess, leaving about 100mm (4in) protruding. Prepare the cores (see page 21) and fit sleeving to each earth core (see Tip).

At roses, feed the cable through the knock-out in the base-plate, and connect the cores to the in-line terminals (precisely which core goes where depends on how the light is wired up and controlled – see pages 37 to 39, 60, and 90 and 91 for details). Sleeve the earth core and connect it to the separate earth terminal.

At switches, the earth core goes to a terminal on the mounting box if this is metal and the switch face-plate is plastic (Fig 50). Live and neutral cores go to the switch terminals (see pages 36, 42 and 43).

At sockets, all cable cores are connected to the face-plate (Fig 51). With metal boxes, a separate earth lead is connected between box and face-plate.

**What you need:**
- side-cutters
- wire strippers
- PVC earth sleeving
- electrical screwdriver
- fixing screws
- small screwdriver

**CHECK**
- that all cores are securely connected to their terminals, by giving each a sharp tug
- that where metal boxes are used and earth cores are connected direct to terminals on accessory face-plates, an extra 'flying earth' is fitted between the face-plate earth terminal and the terminal in the box
- that cable is folded carefully back into mounting boxes as face-plates are fitted

**TIP**
To stop the PVC earth sleeving from slipping off the core while you make the connections, fold the protruding end of the core over it temporarily.

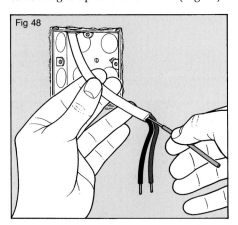

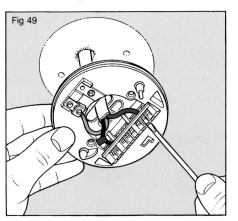

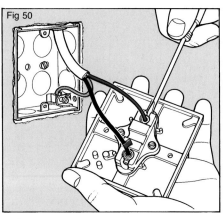

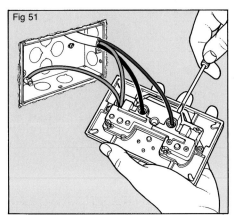

Fig 48 Always cover bare earth cores with green and yellow PVC sleeving.

Fig 49 At roses, wire live and neutral cores to the in-line terminals, and the earth to its terminal.

Fig 50 At switches, live and neutral go to the switch terminals, and earth to the terminal in the box.

Fig 51 At sockets, all cores go to terminals on the face-plate. Note the 'flying earth' link.

# The Importance of Earthing

The concept of earthing can be difficult for the do-it-yourself electrician to understand, but *you ignore it at your peril*; it is one of the most important safety features of any wiring system, protecting you and your family from the risks of receiving an electric shock and also guarding your home against the dangers of fire caused by an electrical fault.

What earthing does is to provide a safe passage to earth for any current that 'strays' from where is should be running – along the live and neutral cores of the house's circuit wiring. Every circuit is connected to earth via an earth continuity conductor, and this must be unbroken; in other words, earth connections must be made at every interruption of the circuit cable, for example, where it is connected to wiring accessories.

All new wiring work should be tested to make sure that the earth continuity is satisfactory, and it is well worth calling in a professional electrician to check your work even if you are confident that you have done a thorough and careful job.

The other vital aspect of earthing is called *cross-bonding*. The purpose of this is to link any exposed metalwork around the house to earth, in case it should come into contact with a live wire and so become live itself. This applies to things like plumbing, heating and gas supply pipework, stainless steel sinks, cast iron baths and central heating controls. In a properly earthed installation, single-core earth cables run from clamps on pipes, sinks, baths and so on to earthing points, either on the house wiring circuits or in the case of mains water and gas supply pipes, to the main earthing terminal at the house's consumer unit or fuse box.

Check that your installation has these cross-bonding links, and that all are undamaged and securely attached. Get them fitted at once if they are missing.

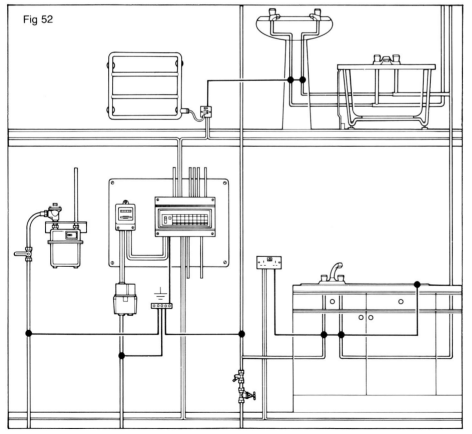

Fig 52

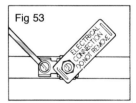

Fig 53

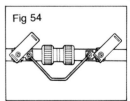

Fig 54

Fig 52  In a properly earthed system, all extraneous metal work is cross-bonded to earth. This includes gas and water supply pipes and metal baths and sinks.

# Good Wiring Practice

The preceding pages have dealt with some of the common (and repetitive) tasks involved in doing your own wiring work. As you will already have realized, most electrical jobs break down into three distinct stages: getting a power supply from A to B, mounting a fitting or wiring accessory at point A and connecting your new work up to the mains at point B. As you tackle the jobs described in the following pages, you should aim to carry out each of these stages as methodically, as carefully and above all as professionally as you can. After all, you have one big advantage over the professionals: your time is your own, and you can afford to spend that bit extra on getting everything absolutely right.

Before you start on your first job, spare a moment to read through this brief check-list of points. They are a summary of good wiring practice, and if you follow them your electrical installation will perform safely and well for the foreseeable future. You will also get the satisfaction of knowing that the job has been well done.

1.  Strip cable and flex sheaths carefully, so you do not damage the insulation on the conductors inside. If you do cut into this, cut off the damaged cores and start again.
2.  Prepare flex and cable cores carefully. Use wire strippers set to match the core diameter, so you do not nick and weaken cable cores or cut off conductor strands from flex cores. If you do this, cut off the cores and start again.
3.  Make connections to terminals with care. Within wiring accessories, make sure that the core is pushed fully into its terminal, and is securely trapped by the terminal screw, which should be fully tightened. Where two or more conductors enter a terminal, twist the cores together with pliers first to prevent one core from pulling out unnoticed.

When wiring up plug terminals, twist the conductor strands neatly together first. With pillar terminals, make sure the core is securely trapped by the retaining screw; with thin cores, double the end over first

to increase the contact area. With stud terminals, wrap the core round the stud clockwise to ensure that it does not unwind as you fit the nut and washer.

With all connections, double-check that the core insulation reaches right up to the terminal so no bare metal is visible.

4.  Always use green and yellow PVC earth sleeving on bare cable earth cores within wiring accessories. When you are wiring up light switches, always put a red 'flag' of PVC tape on neutral cable cores, and on the blue and yellow strapping cores of two-way switching arrangements, to indicate that they are live when the switch is on.
5.  When running cable in walls, always travel vertically or horizontally, never diagonally. Under floors, drill holes in the centre of joists and thread the cable through, rather than cutting notches in the tops, so that the cable is safely out of the reach of screws and nails driven through the floorboards. Always leave some slack on underfloor cable runs, to make future alterations easier to carry out.
6.  When mounting accessories, take care to fit wall mounting boxes squarely – and flush with wall surfaces if they are recessed. With metal flush boxes, always use a rubber grommet to protect the cable sheath from chafing, and fit a flying earth loop between the earth terminal on the box and the one on the face-plate.

Make sure that accessories which are mounted on ceilings are securely attached, either directly to the underside of a joist, or else to a timber batten secured between the joists.

7.  When connecting accessories to their circuit cables, leave enough slack on the cable to allow you to make the terminal connections, then fold the cable back into the mounting box so it is not kinked or trapped by the face-plate. Where you know that two or more cables will be entering a particular mounting box, it pays to fit a box that is deeper than usual at the installation stage to provide extra room for connecting up the various cables within the box.

---

**TURNING OFF THE POWER**
Before you turn to the next section of the book, don't forget the golden rule of electrics:

**make sure the power is switched OFF.**

With consumer units fitted with miniature circuit breakers (MCBs), you do not need to turn the main switch off if you want to work on an individual circuit. Just turn off the individual circuit's MCB.

With fuse-boxes containing rewirable or cartridge fuses, turn off the main switch first, then remove the fuse for the circuit you are working on, and keep it with you until you have finished work.

Always turn off the main switch before you carry out any work within the consumer unit or fuse-box, such as connecting a new circuit cable or rearranging the fuseways.

Beware unexpected links between circuits, such as a socket downstairs wired as a spur from an upstairs power circuit, or two-way switching arrangements between upstairs and downstairs lighting circuits. Even after turning the power to a circuit off, always use a test probe to check that there are no live parts at or near the place where you intend to work.

# INDOOR JOBS

The majority of your wiring projects are likely to involve improving or altering the wiring inside the house, for the simple reason that all your main electrical needs are indoors.

In this section you will find details of how to carry out eighteen different electrical projects, ranging in complexity from fitting a new light switch to wiring a complete new circuit to appliances such as showers and heaters. There is also a summary of what is involved in adding extra circuits if you can no longer meet your needs by extending the ones you have. In each case you will find background information on the job, a checklist of what you need to carry it out, a summary of what the job involves and detailed wiring diagrams to help you make the connections.

## What to Tackle First

Wiring jobs fall into two broad categories – the things you need to do, and the things you want to do. In the former category come jobs such as adding extra socket outlets so you can do away with trailing flexes and overloaded adaptors, while the latter group includes tasks such as fitting new lights to complement a newly redecorated room. Whatever you plan to do, the best way of drawing up a plan of action is to check the wiring facilities in each room of the house in turn. Below are some ideas for improvements you could carry out around the house.

**Living/dining rooms**   Add extra socket outlets for 'leisure' appliances such as TVs, video recorders and hi-fi equipment, to get rid of the spaghetti of flexes and adaptors you have hidden out of sight behind the bookcase.

Fit wall lights, with two-way switching at the room door and by the fittings.

Provide a TV/FM aerial socket outlet, with a concealed coaxial cable link to the roof-mounted aerials.

**Kitchen**   Provide extra socket outlets

for portable appliances at work-top level, perhaps by creating a separate ring circuit for the kitchen.

Wire semi-permanent appliances such as the washing machine, fridge and freezer via fused connection units.

Fit a cooker hood or extractor fan to help remove steam and cooking smells.

Improve the lighting, both at ceiling level and to illuminate the work-tops from beneath the wall cupboards.

**Bathroom**   Wire up a new circuit to supply an instantaneous shower – either over the bath or in its own shower cubicle.

Add a shaver point, perhaps combined with a striplight over the bathroom mirror.

Fit a wall heater and time switch to improve the heating, plus an extractor fan to help combat condensation.

**Bedrooms**   Provide two-way switching for the bedside lights, with ceiling-mounted cord-operated switches.

**Miscellaneous**   Fit a loft light, with a switch on the landing, and convert the hall and landing lights to two-way switching.

Provide telephone sockets at convenient points around the house.

Fig 55 (*above*)   Plan indoor wiring jobs with care so you can carry out the work with the minimum of disruption. It's a good idea to try to co-ordinate the work with redecoration schedules.

### NEW REGULATIONS
From January 1 2005, all new wiring work must meet the requirements of the new Part P of the Building Regulations. With some exceptions, any DIY work must now be tested on completion by an electrician, who will issue a Minor Electrical Installation Works Certificate. In addition, some jobs must be notified in advance to your local authority Building Control Department. Jobs on the following pages are labelled EXEMPT or NOTIFIABLE. *See* page 6 for full details.

# Fitting a Plate-Switch or Dimmer

One of the simplest electrical wiring jobs you can carry out is to replace an existing light switch, either with a new one in a different style (perhaps to match the colour scheme in a newly-decorated room) or with a dimmer switch that will allow you to vary the intensity of the room's lighting from full brightness down to a mere glimmer.

Most switches contain just one cable, linking the switch to the light it controls. One-way switches made of plastic have just two terminals on the back of the face-plate, to accept the live and neutral cable cores (Fig 56). The earth core goes to the terminal in the base of the mounting box. Two-way switches have three terminals on the back; if they are wired for one-way switching, the top terminal and either of the bottom terminals are used to connect the switch cable. All three terminals are used for connecting the switches in two-way switching arrangements (see pages 42 and 43 for more details).

## What to do

Turn off the power at the mains (see page 34). Unscrew the face-plate of the switch you want to replace, and disconnect the switch cable cores from the terminals on the face-plate. Leave the earth connection attached to the mounting box. If more than one cable goes to any terminal, stick a tape tag on each core before you disconnect it and label it to show which terminal it was connected to. Alternatively, make a sketch.

Now reconnect the cores to the new switch, ensuring that you copy the original wiring arrangement precisely. The black core is live when the switch is on, so identify it with a flag of red PVC tape. If you find a bare earth core within the box, disconnect it and cover it with some green and yellow PVC sleeving. Fold the cable back neatly and secure the face-plate to the box. To fit a dimmer switch, follow the wiring instructions included with it.

**What you need:**
- screwdriver
- electrical screwdriver
- red PVC tape
- green and yellow earth sleeving

**CHECK**
- that core conductors are securely held within the terminals, with no bare metal showing
- that terminal screws are tight
- that neutral cores are identified as live with a flag of red tape
- that earth cores are sleeved

**TIP**
Keep the old face-plate screws to secure the new switch to the box. The new screws may have a different thread pattern to that in the lugs on the old box, and so may not engage properly.

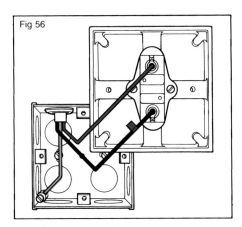

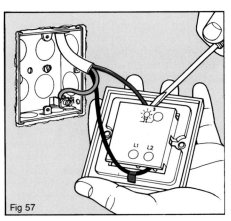

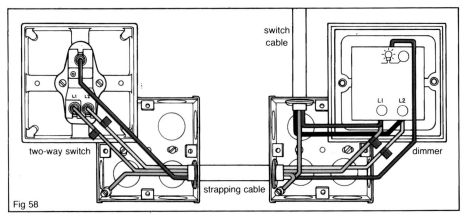

Fig 56   How the switch cable cores are connected to a typical plastic one-way plate-switch.

Fig 57   To fit a new dimmer switch, simply follow the wiring instructions included with it.

Fig 58   Two-way switches have three terminals, and are linked with special three-core cable. You can replace either of the two-way switches with a dimmer switch; here the dimmer replaces the 'master' switch. *See* pages 42 and 43 for more details.

# Installing a New Light Fitting

Many homes have just a traditional ceiling rose and pendant lamp-holder in the centre of each room, and even new homes seldom have anything more adventurous. You can of course ring the changes with fittings like these, by using lampshades in different shapes and styles. However, you may prefer to replace the rose and lamp-holder with a different type of light fitting altogether – this is a job which will involve making some simple alterations to the existing wiring.

Most ceiling-mounted light fittings are designed to be screwed directly to the ceiling, and come with a short length of flex attached, ready for connection to the circuit wiring. Since such connections have to be made within an enclosure, you need to provide a mounting box recessed into the ceiling, over which the base of the new light fitting can be mounted. In many cases, the fitting is screwed directly to the box; if not, it will need the support of a batten fixed between the joists in the ceiling void above.

## What to do

Turn off the power at the mains (see page 34). Unscrew the cover of the ceiling rose you aim to replace, and disconnect the pendant flex from its terminals.

Next, label the individual cable cores with tape tags, or make a sketch of the wiring arrangement (see pages 92 and 93 for reference diagrams of wiring at ceiling roses). Disconnect the cores and remove the rose. Cut a hole in the ceiling, secure a batten between the joists above and fit the mounting box (see page 31).

How you connect the new fitting to the circuit cables depends on how the rose was wired up. Loop-in roses may have two, three or four cables present, while a junction-box rose has only one. Connect the flex and cable cores as shown in (Fig 61) for loop-in wiring, or (Fig 62) for junction-box wiring, using strip connectors. Push the connectors neatly into the mounting box and secure the new fitting in position.

**What you need:**
- new light fitting
- round mounting box
- batten and scrap wood
- strip connectors
- red PVC tape
- earth sleeving
- screwdriver
- electrical screwdriver
- padsaw
- wood screws

**TIP**
Some light fittings are designed to be screwed directly to the lugs in the mounting box using machine screws (Fig 63).

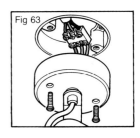
Fig 63

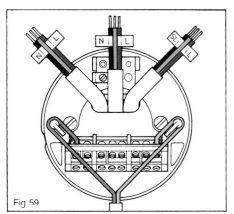
Fig 59

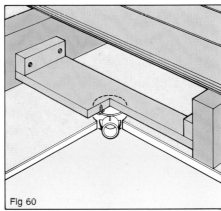
Fig 60

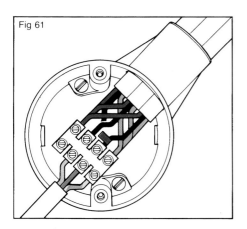
Fig 61

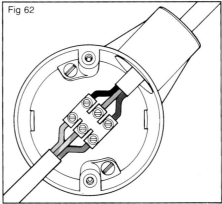
Fig 62

Fig 59   Label the cable cores at the old rose to show which terminal they were connected to.

Fig 60   Fit a mounting box on a batten, and feed in the circuit cable(s).

Fig 61   With loop-in wiring all red cores go to one terminal, all earth cores to another. The live flex core is linked to the switch neutral core (with the red flag), and the neutral flex core to the black cores.

Fig 62   With junction-box wiring the cable and flex cores are linked like to like.

# Extending a Lighting Circuit

Instead of merely installing new light fittings at the positions of existing ceiling roses, you may want to be a little more adventurous and add extra lights elsewhere around the house.

You can connect into an existing lighting circuit at any convenient point to provide a power supply for a new light, subject to one proviso. Lighting circuits in the home are generally rated at 5 amps, which means that they can supply a maximum of 1,200 watts of power. Each lighting point is assumed to use an average of 100 watts, so in theory each circuit can supply twelve lights. In practice, allowance is made for the fact that some light fittings containing several lamps may consume more than 100 watts, so to allow for this each circuit is in practice restricted to supplying eight lights. It is therefore vital to check how many existing lights a circuit supplies before planning to add extra lights to it, to avoid the risk of overloading.

## Wiring Options

Irrespective of how your lighting circuits are wired up, you can provide power for extra lights in one of two main ways. The first is to cut into the main circuit cable and to wire in a four-terminal junction box (Fig 64, below). Within this box the cut ends of the circuit cable are linked, and cables are also wired in to supply the new light and to run to its switch (Fig 66).

The second option is again to cut the circuit cable, but this time to wire in a new loop-in rose, with a new cable running to its own switch (Fig 65, below and Fig 68, page 39).

## What to do

**Option 1** If you choose to wire in a junction box, first turn the power off at the mains (see page 34). Lift floorboards or remove loft insulation, depending on where the lighting circuit cables run, so you can locate the most convenient point at which to make your connection. The junction box can be sited anywhere within reason but must be securely mounted to a joist, so look for a cable running close to a joist with enough slack in it to allow an easy connection to be made. Double-check that it is a circuit cable, and not a switch cable: you cannot take power from one of these. Cut the cable and prepare the cores (see page 21).

Now run new cables from the junction box to the new light and switch positions (see pages 24 to 27). Within the junction box, connect the circuit and switch cable live cores to one terminal, the switch cable neutral and light cable live to the second, the remaining neutrals to the third and the sleeved earths to the fourth.

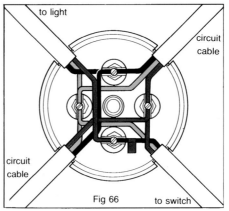

Fig 66

to light

circuit cable

circuit cable

to switch

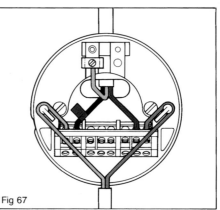

Fig 67

**What you need:**
- side-cutters
- wire strippers
- 5-amp 4-terminal junction box
- screwdriver
- wood screws
- red PVC tape
- earth sleeving
- electrical screwdriver
- 1mm$^2$ two-core and earth PVC-sheathed cable
- new rose or light fitting
- new plate-switch
- general building and woodworking tools

**CHECK**
- that the circuit you plan to extend has enough spare capacity
- that the cable from which you intend to take power is a main circuit cable and not a switch cable. You cannot take power from the latter, since it carries current only when the switch is turned on

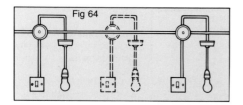

Fig 64

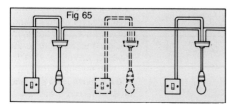

Fig 65

Fig 64   Cut in a new four-terminal junction box.

Fig 65   Alternatively, fit a new loop-in rose.

Fig 66   Connecting the existing circuit cables and the new cables to light and switch within a 4-terminal junction box.

Fig 67   Connecting the new cable from the junction box to its new ceiling rose.

# Extending a Lighting Circuit

**Option 2** If you decide to wire in a new loop-in rose, your first step is to locate a suitable circuit cable to supply the new rose as you do when cutting in a junction box. Since the cable will be connected directly into the new rose, the siting of the rose is obviously restricted by the positions of existing cables in ceiling voids. Where no cable runs close to the proposed position of the new light, you will have the use a four-terminal junction box instead (see page 38).

If you do find a suitable cable, turn off the power at the mains (see page 34) and cut the cable as close as possible to the new rose position. Prepare the circuit cable cores, and run in the new switch cable.

Within the rose (Fig 69), connect all the live cores to the central bank of terminals (often marked 'line loop-in'), the switch neutral core to the switch live terminal (marked 'sw'), the remaining neutrals to the neutral terminal and the sleeved earths to the separate earth terminal on the base of the rose.

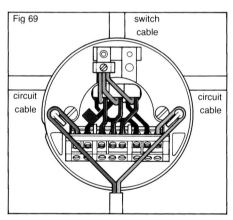

## Plug-in ceiling roses

The permanent connection to the mains of pendant lamp-holders and other suspended fittings has long been a nuisance, making it difficult to take fittings down for cleaning or replacement, or during ceiling decoration. Plug-in ceiling roses, called luminaire support couplers (LSCs), allow fittings to be connected and disconnected as easily as plugging a portable appliance into a socket outlet. The base of the rose is mounted on the ceiling just like an ordinary rose, and is wired up in a similar way.

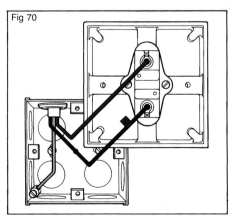

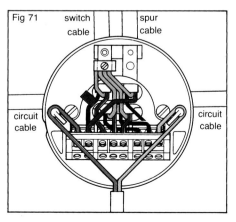

Fig 68 (*above*) Plug-in ceiling roses allow easy removal of fittings for cleaning or replacement.

Fig 69 How the existing circuit cables and the new switch cable are connected within the new loop-in rose.

Fig 70 How the new switch is wired up, whether run from a junction box or a loop-in rose.

Fig 71 If you want to add a new light that is to be controlled from an existing switch, simply run a spur cable from an existing loop-in rose to the new rose, and wire it up as in Fig 67. At the existing rose, wire the spur cable cores to the live and neutral flex core terminals as shown here.

# Installing Wall Lights

Ceiling-mounted lights are fairly good at providing general lighting, but because the light source is usually in the middle of the room they are not as effective for background lighting, for highlighting features of the room or for providing task lighting for things like reading. Lights mounted on the room's walls can do all these things, as well as adding to the decor. They come in a huge range of styles, and can provide anything from a diffuse glow to a brilliant spotlight.

Once you have chosen the type of light you want, you have several decisions to make – where to site them, how to provide a power supply, how to switch them on and off, and how to conceal the wiring. This last point is the most difficult, since unless you can live with surface-mounted wiring, you are doomed to ruin your wall decorations. The best compromise is to disguise surface-run cables with paint to match the colour scheme until they can be chased in later on.

## What to do

The main decision concerns how to supply power to the lights, and here you have several options.

**Option 1**  Connect the supply cable as a spur to an existing loop-in ceiling rose (Fig 72) or junction box (Fig 73). In each case the new wall light will be controlled by the existing switch, unless it has its own integral push-button or pull-cord switch. The spur cable can feed more than one wall light, so long as this will not overload the lighting circuit concerned. In this case the spur cable is run to a three-terminal junction box, and each light is connected to the box by its own cable with like cores being connected into like terminals.

**Option 2**  Do away with an existing ceiling-mounted light altogether, and use its power supply for your new wall lights. Disconnect the existing light's supply cable(s) and remove it/them. Then draw

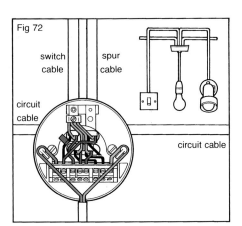

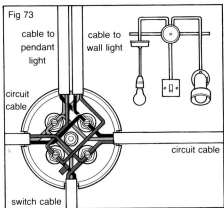

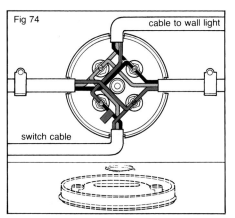

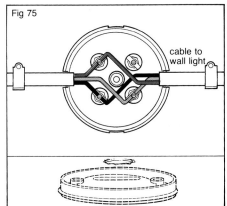

Fig 72  How the spur cable is wired into an existing rose (Option 1).

Fig 73  How the spur cable is wired into an existing junction box (Option 1).

Fig 74  How a junction box is wired up to replace an existing loop-in rose or light (Option 2).

Fig 75  How a junction box is wired up to replace a rose wired from its own junction box (Option 2).

# Installing Wall Lights

the cable(s) up into the ceiling void and reconnect it/them to a four-terminal junction box; use all four terminals (Fig 74) where two or more cables are present, and just three (Fig 75) if there is only one cable. Run a spur cable on from the box to the new wall light(s), which will be controlled by the same switch as the existing light.

**Option 3** Make a brand new connection to an existing lighting circuit at a convenient point, using a four-terminal junction box in the same way as if you were to extend a circuit to supply a new rose (see page 38). This has the advantage of allowing the new lights to have their own independent switch. Run new cables from the box to the light(s) and to the new switch (Fig 77).

**Option 4** Take the power supply as a spur from a power circuit (see pages 46 and 47 for more details). The spur runs to a fused connection unit (FCU), which can also act as the light's on/off switch, and then on to the new light (Fig 78).

Fig 76 (*above*) Wall lights help add a new dimension to any room's lighting.

```
CHECK
•  that you don't
   overload lighting
   circuits. Each can
   supply a maximum of
   eight lights only
```

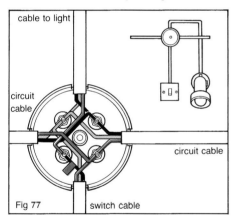

cable to light

circuit cable

circuit cable

Fig 77          switch cable

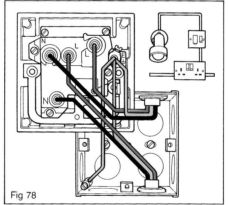

Fig 78

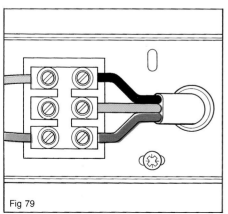

Fig 79

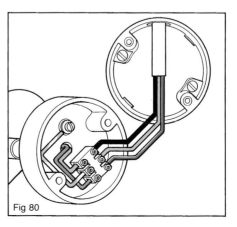

Fig 80

Fig 77   How a junction box is wired in to an existing lighting circuit (Option 3).

Fig 78   How a fused connection unit is wired to supply a new light (Option 4).

Fig 79   How the supply cable is connected to a light with a terminal block.

Fig 80   How the cable is linked to a light with flex tails, mounted over a conduit box.

# Providing Two-Way Switching

Each light around the home is usually controlled from just one switch, which makes or breaks the circuit to the light as it is operated. However, there are a number of situations where it would be more convenient to be able to switch a particular light on or off from more than one position. Examples include wall lights in bedrooms, switched by the bed as well as by the door, and also lights in the hall and on the landing controlled from either upstairs or downstairs.

This arrangement is known as two-way switching, and involves using two switches connected with special cable containing three colour-coded cores plus an earth. Each switch must be suitable for two-way switching, which means there must be three terminals on the back of each face-plate.

You can add more switches to the system if you want to provide more control points, by using special switches with four terminals on the back called intermediate switches (see page 43).

## What to do

Turn off the power at the mains (see page 34). Unscrew the face-plate of the switch that controls the light at present, and disconnect the cable. If this is run in a conduit you may be able to feed the new three-core cable in alongside it; otherwise cut a chase for it up to the ceiling, parallel to the existing switch cable.

Run this three-core cable through the ceiling void to a point above the new switch position. Then cut a chase for it down the wall, run in the cable and fit the mounting box for the new switch.

You can now fit a new two-way switch at each switch position. At the original switch position, connect the original two-core switch cable so that its red and black cores go to the bottom pair of terminals (usually marked L1 and L2).

Next, connect the blue core of the three-core cable to the L1 terminal at each switch, the yellow core to the L2 terminals and the red core to the top terminal (usually marked C, or common). Add earth links between face-plate and mounting box if the switches are metal. Then fit the face-plates to the mounting boxes, restore the power and test the system.

Fig 81　Two-way switching gives you the convenience of being able to control bedside lights from the bed as well as by the door.

> **What you need:**
> - 2 new two-way switches
> - mounting box
> - 1mm$^2$ three-core and earth cable
> - red PVC tape
> - earth sleeving
> - conduit (optional) for new switch cable run
> - side-cutters
> - wire strippers
> - screwdriver
> - electrical screwdriver
> - wood screws
> - wall-plugs
> - filler
> - general building and carpentry tools

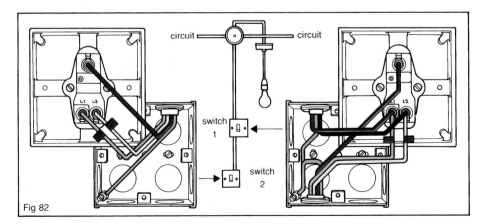

circuit　circuit

switch 1

switch 2

Fig 82

Fig 82　How the new two-way switches are wired up at the existing and new switch positions.

# Providing Two-Way Switching

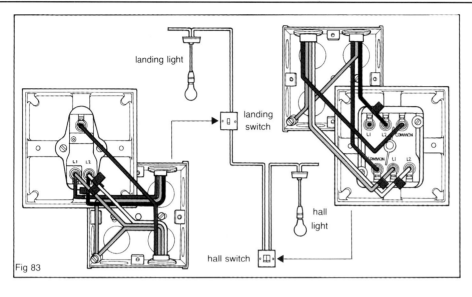

Fig 83   In stairwells you can provide partial two-way switching as shown here. The one-way landing light switch is replaced by a one-gang two-way switch, and the existing one-gang one-way hall light switch is replaced by a two-gang two-way switch. The two are linked with three-core cable. The hall light is still controlled just from the hall, but the landing light can be switched on or off from either switch position.

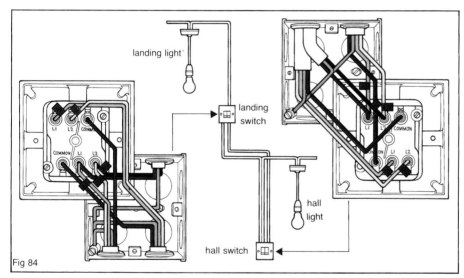

Fig 84   If you want full two-way switching, with both lights controllable from upstairs and downstairs, replace both switches with two-gang two-way switches, and link the switches with two separate three-core cables. Connect the cores as shown here.

## CHECK
- that all cores are flagged with red PVC tape to indicate that they are live
- when linking hall and landing lights that the power to both upstairs and downstairs lighting circuits is off before you start work

## Adding more Switches

If you want more than two switches to control a particular light, you can extend the two-way switching arrangement to a multi-way set-up by using extra switches called intermediate switches. These are fitted, as the name suggests, between the first and last switches, which are both the two-way type. Three-core cable is again used to link the switches.

Intermediate switches have two pairs of terminals which are used to connect in the blue and yellow cable cores. The red cores are linked via a strip connector (Fig 85).

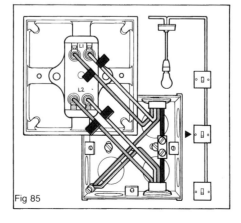

Fig 85   How the cores are connected within an intermediate switch. Note that the red cores are linked within the mounting box using a strip connector.

43

# Converting Single Sockets to Doubles

The more electrical appliances we buy, the more socket outlets we need to plug them into, and few homes seem to have enough, especially in rooms like kitchens and living rooms where a lot of appliances are used. One way of increasing the number of sockets available without the need for any complex wiring work is to convert existing single sockets to double (or even triple) ones. There are no restrictions on the number of sockets a modern power circuit can supply, so you can happily convert all your sockets in this way. However, you should *not* convert old-style sockets taking round-pin plugs unless you have the existing circuit wiring checked first by a professional electrician.

As far as the actual conversion job is concerned, what you do depends on whether your existing socket is flush- or surface-mounted, and how you want the new one to be mounted. There are four options which involve differing amounts of work.

## What to do

Turn off the power at the mains (*see* page 34).

**Option 1** If your existing socket is surface-mounted, the simplest conversion is to remove the existing single face-plate and mounting box, and replace it with a double-surface mounting box (Fig 87). First remove a knock-out from the new box, feed in the cables and reconnect them to the terminals on the new double-socket face-plate. Then attach the face-plate to the box.

**Option 2** Convert an existing flush single socket to a surface-mounted double one. Again unscrew and disconnect the existing single-socket face-plate. Then mount a new double-surface mounting box over the old flush box (Fig 88). Most double boxes have a pair of fixing holes in the base at 60.3mm centres, to match the distance between the lugs on single flush boxes. Note that you must use the machine screws that held the old socket face-plate

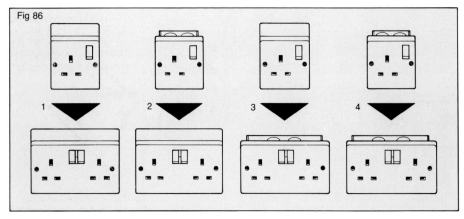

Fig 86

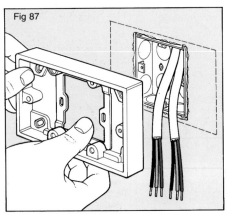

Fig 87

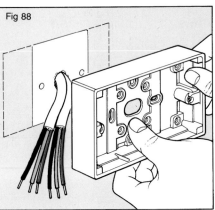

Fig 88

Fig 86 Four conversion options are possible: surface-to-surface (1), flush-to-surface (2) surface-to-flush (3) or flush-to-flush (4).

Fig 87 For Option 1, remove the old socket and box and fit a new larger box in its place.

Fig 88 For Option 2, remove the old socket face-plate and mount a new larger box over the existing one. Use the old retaining screws to secure the new box to the lugs on the old one.

# Converting Single Sockets to Doubles

to its box to fix the new mounting box in place. Then feed in the cables and connect them to the face-plate as discussed in the first option.

**Option 3**   Convert an existing surface-mounted socket to a flush one. Remove the old face-plate and its mounting box. Mark the outline of the new flush double-mounting box on the wall. On solid masonry walls, chop out a recess for the new box (see page 29), taking care not to damage the cables as you do so. Then feed the cables into the new box through a grommet, secure the box in the recess, make good round it and connect up the new double face-plate. On hollow walls, make a cut-out in the plasterboard and fit a special flush-mounting box in the hole (see page 30 for more details).

**Option 4**   Convert an existing flush single socket to a flush double one. Here you have to enlarge the existing recess to take a double box. This involves chopping out solid walls (Figs 90 and 91) or making a bigger cut-out in hollow ones (Figs 92 and 93)

Fig 89 (*above*)
Converting single sockets allows you to plug in extra appliances without having to do any complex wiring work.

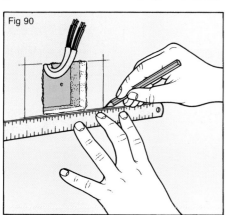

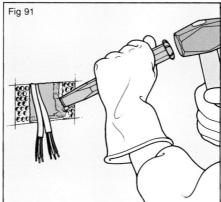

Fig 90   For Option 4, remove the old socket and box and mark the outline of the new box on the wall.

Fig 91   Then honeycomb the masonry and chop it out to create a new recess for the double box. Take care not to damage the circuit cables.

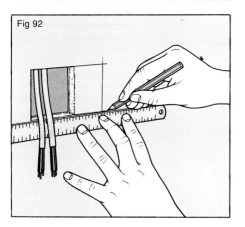

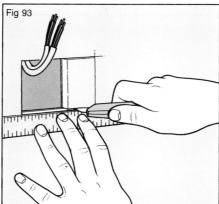

Fig 92   On hollow walls, mark the outline of the new box on the wall.

Fig 93   Then cut away the plasterboard to form a larger cut-out, and fit a special flush box.

# Adding Extra Sockets

You may find that converting single sockets to doubles still does not provide enough positions to plug things in, or that you simply need sockets where none are at present fitted. The answer is to add extra sockets to your circuits.

As already mentioned, there are no restrictions to the number of sockets that a modern power circuit can supply. The only limits imposed by the wiring regulations concern the total floor area of the rooms each circuit serves (see pages 88 and 89), so you need to take care not to exceed these by adding sockets supplied by one circuit but located in rooms otherwise served by a different circuit.

You can add sockets to circuits in three different ways – as spurs connected to existing sockets on the circuit, as spurs connected to three-terminal junction boxes cut into the circuit wiring, or as direct additions to the circuit wiring if the way the cable is run will allow them (see also page 47 for details).

## Choosing a Suitable Socket

You can wire your spur cable only to a socket on the main circuit. You are not allowed to use a socket that is on a spur (Fig 94, top), or one on the main circuit that already supplies a spur (Fig 94, middle). Under current wiring regulations each spur may serve only one socket (although this can be either a single or a double socket). Formerly, two separate sockets could be served by one spur cable. The intermediate socket on such a spur (Fig 94, bottom), contains two cables and might appear to be on the main circuit, but must not be used to wire in a spur.

## What to do

**Option 1** The simplest way of adding new sockets to the existing wiring is to connect its supply cable at an existing socket (see the paragraph below on choosing a suitable socket), and run it as a sort of branch line to the new socket position.

Decide where you want your new socket to be, and locate a suitable socket to which you will connect the spur cable. Then fit the new mounting box (see pages 28 to 30), connect the spur cable to the new socket face-plate (Fig 95), fit it to its box and run the spur cable back to the existing socket (see pages 23 to 27).

Now turn off the power at the mains (see page 34). Unscrew the face-plate of the socket you are going to use for the spur connection, feed in the spur cable and connect it to the terminals (Fig 96). Carefully fold the three cables back into the mounting box, and replace the face-plate. Turn on the power and test the system.

### What you need:
- new socket
- new mounting box
- 2.5mm² two-core and earth cable
- earth sleeving
- side-cutters
- wire strippers
- screwdriver
- electrical screwdriver
- wood screws
- wall-plugs
- electric drill plus twist/masonry bits
- general building and carpentry tools

### CHECK
- that sockets containing two cables and fed by ring mains are on the main circuit. With the power *off*, disconnect the live cores from their terminal and link them with a continuity tester. It will light up if the socket is on the ring, but not if it is on a spur.

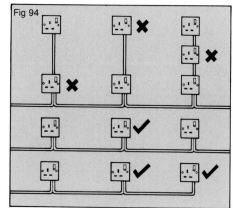

Fig 94

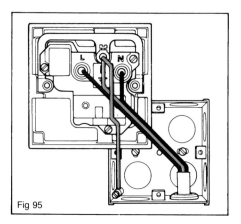

Fig 95

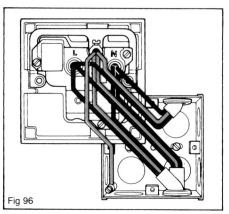

Fig 96

Fig 94 On ring circuits, sockets on spurs or supplying spurs cannot be used as connection points (top), but any socket actually on the ring can be used (centre). On radial circuits any socket can supply a spur (bottom).

Fig 95 How the spur cable is connected to the new socket face-plate. Note the earth link between face-plate and box.

Fig 96 How the spur cable is wired into the socket on the main circuit.

# Adding Extra Sockets

**Option 2** If connecting your spur to an existing socket is inconvenient for any reason – for example, it might involve an awkward cable route – an alternative is to connect the spur cable directly to the main circuit cable using a 30-amp three-terminal junction box. Round ones in brown plastic (Fig 98) are the norm, designed to be hidden beneath the floorboards, but if you have surface wiring you can use a smart surface-mounted white rectangular type instead (Fig 99).

Wire up the spur as described on page 46. Then locate the main circuit cable at an easily accessible point beneath the floorboards, and check it has enough slack on it to allow you to make the connections. Turn off the power at the mains (see page 34), cut the cable, prepare the cores and connect them to the terminals. Screw the box to the side of a joist (Fig 98) or to the skirting board (Fig 99), depending on which type of box you are installing, then feed in the spur cable and connect it up as shown.

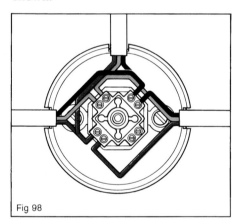

Fig 98

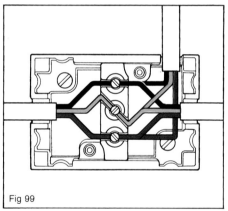

Fig 99

Fig 97 (*above*) Adding extra sockets provides all the power points you need.

Fig 98 How the circuit and spur cables are wired into a 30-amp round junction box.

Fig 99 How the cables are wired into a rectangular surface-mounted box.

**Option 3** This option can be used only if existing circuit cables are lying loose in floor voids and have enough slack to allow them to be drawn up behind the skirting board to the new socket position. If this is the case, the cable can be cut and connected directly to the terminals on the new socket face-plate (Fig 101). Cut a channel for the cable behind the skirting board, feed it up as a loop and then turn the power off at the mains (see page 34) before cutting the cable, preparing the cores and making the connections. Take care when drawing the cable loop up not to overstress it and risk damaging the cores inside.

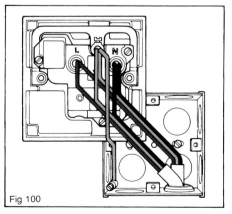

Fig 100

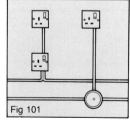

Fig 101

Fig 100 (*above*) Use junction-box spurs (right) if there is no convenient socket.

Fig 101 How circuit cables are direct-wired to a new socket.

# Using Fused Connection Units

Socket outlets are the ideal source of power for portable appliances such as kettles, toasters, food mixers and hair driers, or for 'movable' equipment such as televisions, hi-fi, table lamps and so on. However, many appliances that need a power supply – things like washing machines, fridges, freezers and waste-disposal units – are essentially fixed, and need to remain permanently connected to the mains supply. For appliances like these, it is more convenient to make the connection via a fused connection unit (FCU) than to use a plug and socket.

The advantages of an FCU over a plug and socket are that it is dedicated to a particular appliance and provides a permanent connection to the mains, so the appliance cannot be disconnected inadvertently (essential for food freezers, for example). An FCU can also be used to provide a sub-circuit to things like wall lights. The unit contains a cartridge fuse, identical to those used in plugs.

## Choosing the right FCU

FCUs are the same size as a single socket outlet, and are fitted over standard single mounting boxes. There are several versions, with or without switches and indicator lights, and with the option of flex entry through the front or edge of the face-plate for use with appliances. A different version is used for wiring fused spurs – to lights, for example – which allows cable to be connected instead of flex. The unit has a double-pole switch, and its fuse is housed in a recessed carrier accessible from the front of the face-plate. Fit a fuse rated to match the appliance wattage.

## What to do

An FCU is just a special sort of socket outlet, and is connected to a power circuit in exactly the same way, either on the main circuit or as a spur.

Start by deciding where the FCU is to be positioned. For floor-standing or fixed kitchen appliances, for example, it is often best to site the FCU on the wall at the back of the recess the appliance will occupy and below work-top level so that the flex is out of sight when connected (Fig 102).

Install the unit's mounting box, and run the supply cable(s) to the box, ready for connection to the face-plate. Connect the supply cable cores to the terminals marked 'feed'. If a flex connection is to be made, feed the flex through the face-plate and connect its cores to the terminals marked 'load' (Fig 103). Then anchor its sheath in the cord grip. If an outgoing cable is being connected, simply connect its cores to the 'load' terminals (Fig 104).

**What you need:**
- fused connection unit
- mounting box
- 2.5mm² two-core and earth supply cable
- earth sleeving
- 3-, 5- or 13-amp fuse to suit appliance or sub-circuit rating
- side-cutters
- wire strippers
- screwdriver
- electrical screwdriver
- electric drill plus masonry bit
- general building and carpentry tools

**CHECK**
- that the fuse in the FCU is the appropriate rating for the job – 3-amp for appliances rated at up to 720 watts, 13-amp for more powerful appliances and 5-amp for lighting sub-circuits
- that the flex sheath is held securely in the cord grip inside the FCU

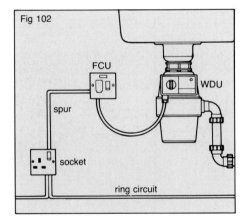

Fig 102

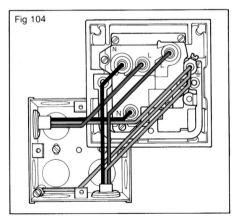

Fig 103 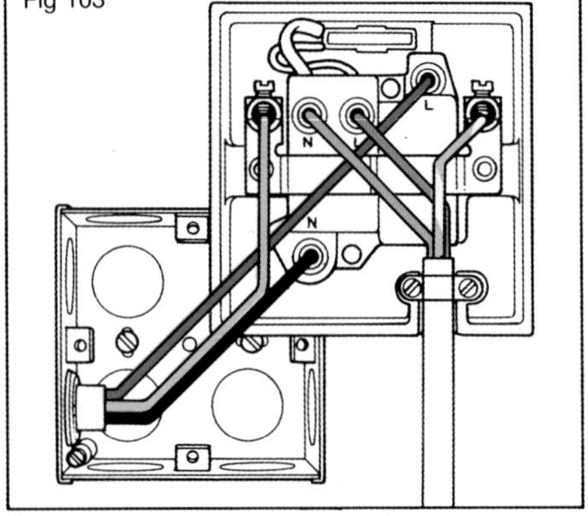                    Fig 104

Fig 102  A typical installation to a fixed appliance, with the FCU wired as a spur from a power circuit.

Fig 103  How the supply cable and incoming flex are wired to the terminals of a flex-entry FCU.

Fig 104  How the incoming and outgoing cables are wired to the terminals of a cable-entry FCU to supply a fused spur.

# Wiring Extractor Fans and Cooker Hoods

Extractor fans are a very useful aid in combating the effects of condensation in the home, especially in traditionally steamy rooms such as kitchens and bathrooms. Cooker hoods, so long as they are extractors and not just filters and recirculators, perform the same function. Both need a permanent connection to the mains supply, but how you do this depends on the type of fan you are installing and how you want to control it.

If the fan does not have its own pull-cord switch, the simplest way of wiring it is to run a fused spur from a nearby power circuit (Fig 109) to a switched FCU, and then to connect the flex from the fan into the FCU (Fig 105). This will also act as its on/off switch.

In enclosed rooms with no window, the fan is essential for ventilation, and here it is better to link the fan to the room's lighting circuit so the fan operates whenever the light is turned on. A spur cable is run from the light itself if this is wired on the loop-in principle (Figs 106 and 110), or from its junction box otherwise (Fig 107 and 111). In each case the spur is run to an FCU, which allows the fan to be isolated from the mains for cleaning, and to be switched off if required while the room light is on.

If the fan does have a pull-cord switch, you can wire it as a simple spur from a nearby lighting circuit, using a three-terminal junction box to make the connection from the fan to the mains (Figs 108 and 112).

## What to do

**Option 1**   Run a spur to a conveniently-sited FCU and connect in the fan's flex.

**Option 2**   Run a spur from a loop-in light to an FCU and connect the flex as in A.

**Option 3**   Run a spur from the light's junction box to an FCU wired as in A.

**Option 4**   Connect the fan to a lighting circuit via a three-terminal junction box.

What you need:
- fan or cooker hood
- switched FCU
- 1mm$^2$ two-core and earth cable
- earth sleeving
- 3-terminal junction box
- electrical tools

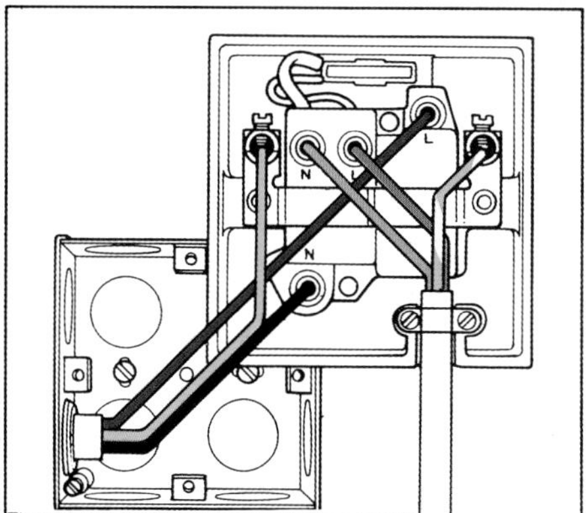

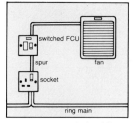

Fig 109   Unswitched fan on power circuit spur.

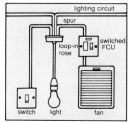

Fig 110   Unswitched fan wired to room light fitting.

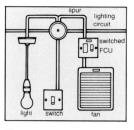

Fig 111   Unswitched fan wired to light's junction box.

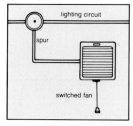

Fig 112   Switched fan wired to 3-terminal junction box.

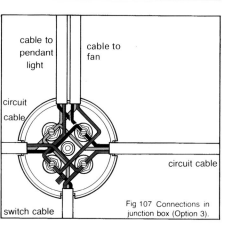

Fig 105 Connections in FCU (Options 1 to 3).

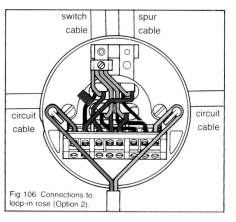

Fig 106 Connections to loop-in rose (Option 2).

Fig 107 Connections in junction box (Option 3).

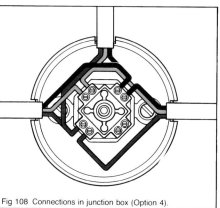

Fig 108 Connections in junction box (Option 4).

49

# Wiring Shaver Units

Electric shavers can be plugged into a standard socket outlet via a special shaver adaptor, but this is often inconvenient because most sockets are at skirting-board level. It makes more sense, therefore, to provide special sockets just for shavers at a convenient height above floor level where they are most likely to be used – generally in the bathroom or the bedroom. Both can be wired up as spurs from either a lighting or a power circuit.

In bathrooms, washrooms and cloakrooms, electrical safety is of paramount importance, and so shaver supply units containing an isolating transformer must always be fitted. These provide an earth-free supply, so the user is completely isolated from the main, and contain a cut-out to prevent any other appliance being plugged into them. They generally offer two output voltages – 110V and 240V – so they can be used by visitors from countries with mains voltage lower than in the UK; you select the voltage by means of a switch, or by plugging into two of the three sockets

on the unit. This type of unit is also available combined with a striplight.

In bedrooms and other locations safely away from a water supply, you can use the smaller shaver socket outlet, which does not contain a transformer but has a low-amperage fuse to prevent other appliances being plugged into it.

## What to do

**Option 1** Run a spur cable from a three-terminal junction box (Fig 108) to the shaver unit, and connect in the cable (Figs 113 to 115).

**Option 2** Connect the spur cable into the room's light fitting (Fig 106), or into the junction box supplying it (Fig 107).

**Option 3** Run a fused spur in 2.5mm$^2$ cable from a power circuit to a fused connection unit (Fig 78), and then run 1mm$^2$ cable from there to the shaver unit.

**Option 4** As Option 3, but with the spur run from a junction box on a power circuit.

What you need:
- shaver supply unit *or* shaver socket
- mounting box
- 1mm$^2$ or 2.5mm$^2$ two-core and earth cable
- earth sleeving
- FCU or junction box
- electrical tools

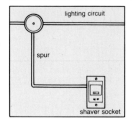

Fig 116  Wired to junction box on lighting circuit.

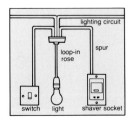

Fig 117  Wired direct to room's loop-in light.

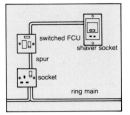

Fig 118  Wired on fused spur from socket outlet.

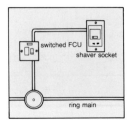

Fig 119  Wired on fused spur from junction box.

Fig 113  Shaver supply unit.

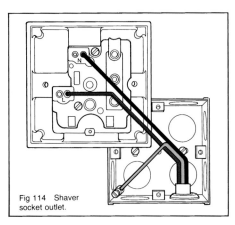

Fig 114  Shaver socket outlet.

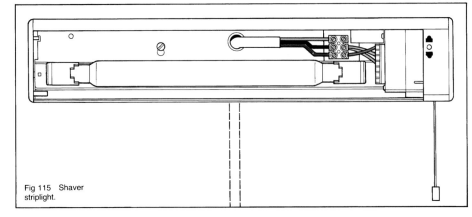

Fig 115  Shaver striplight.

# Wiring Central Heating Controls

If you have a central heating system, it will be operated by a variety of controls, and these take their power supply from the house wiring via a fused connection unit. From the FCU a spur cable is run to a special junction box, to which the system's main controls are wired up.

The main control is the programmer, which controls the periods when space and/or water heating are supplied. The room thermostat switches the space heating on and off as it senses air temperature changes. The cylinder thermostat (if fitted) monitors the water temperature in the hot cylinder, and activates a motorized valve to divert water between the heating and hot water circuits as required.

The wiring between the various control components and the main junction box is run using special multi-core flex. The diagrams below are examples only; always follow the installation instructions supplied with your controls when wiring up your own system.

## What to do

Begin by deciding on the optimum siting for the programmer and room thermostat. The programmer should be sited wherever is most convenient for access to the controls – the kitchen, the hall or the living room are all suitable locations. The room thermostat must be positioned out of draughts and away from heat sources such as radiators. Since you probably spend most of your time in your living room, place it there so that room is always at the design temperature. Fit the cylinder stat to the hot cylinder, and position motorized valves as required by the system design.

Connect each control to the relevant terminals in the main junction box, as dictated by the wiring diagrams supplied with the controls. Run 1mm² two-core and earth cable from the box to a fused connection unit fitted with a 3-amp fuse, and then connect the FCU as a spur to a nearby power or lighting circuit.

**What you need:**
- fused connection unit
- mounting box
- main junction box
- controls
- 1mm² two-core and earth cable
- earth sleeving
- multi-core flex
- conduit box for valve connections
- electrical tools

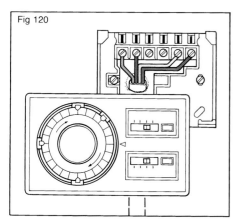

Fig 120

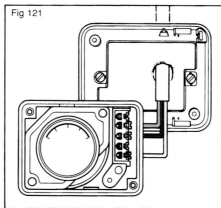

Fig 121

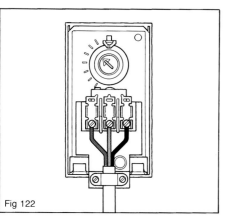

Fig 122

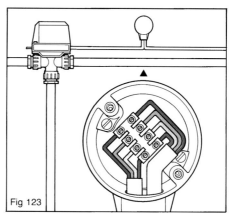

Fig 123

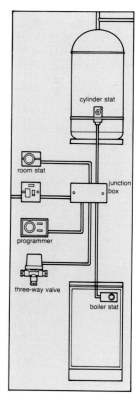

Fig 124 The controls are linked to a junction box supplied by an FCU.

Fig 120 System programmer.

Fig 121 Room thermostat.

Fig 122 Cylinder stat.

Fig 123 Connections between flex tail of motorized valve and flex to main junction box.

# Wiring Wall Heaters and Towel Rails

Most portable heating appliances – radiant fires, fan convectors and the like – are usually plugged into a convenient socket outlet in the room where they are needed. However, some heaters are designed for permanent wall mounting, and it makes sense for these to be permanently connected to the mains too. In bathrooms, where ordinary socket outlets are not allowed, a permanent connection is required by the wiring regulations in any case, and unless the heater has its own cord-operated on/off switch you will also have to provide a separate pull switch to control it if the switch is situated within 2m (6ft 6in) of the bath or shower.

A heater with its own switch is wired up by connecting the flex into a switched fused connection unit (Fig 125, top). If you need a separate ceiling switch, fit an unswitched FCU on the supply spur, and run cable from the switch to a flex outlet plate (FOP) (Fig 125, bottom). The heater flex is then wired to this.

## What to do

Start by mounting the heater on the wall, then select the appropriate wiring option for switched or unswitched heaters. In either case, make sure the cartridge fuse in the FCU is rated at 13 amps.

**Option 1** For switched heaters, use a switched fused connection unit with flex outlet (Fig 126). Mount it close to the heater, connect the live and neutral flex cores to the 'load' terminals and the earth core to the earth terminal. Run cable from the FCU's 'feed' terminals back to your chosen connection point on the main power circuit, turn off the power at the mains and make the connection.

**Option 2** For unswitched heaters, fit a flex outlet plate near the heater, wire in the flex (Fig 127) and run cable back from it via a 30-amp double-pole ceiling switch (Fig 128) to an unswitched FCU. From there, make the connection to the main power circuit as before.

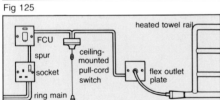

Fig 125

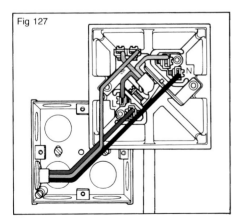

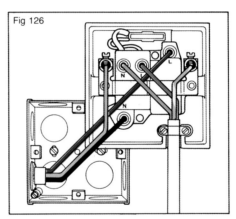

Fig 126

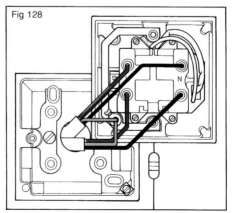

Fig 127

Fig 128

Fig 125 Circuit layouts for wiring switched (top) or unswitched heaters (bottom).

Fig 126 How the wiring is connected within an FCU.

Fig 127 How the wiring is connected within a flex outlet plate.

Fig 128 How the wiring is connected within a ceiling switch.

# Wiring Immersion Heaters

Immersion heaters are widely used for providing stored hot water in homes without a boiler or heating system, and are generally also fitted as a back-up to hot water systems which are normally centrally heated. In the latter case they can also be used in summer, when it is uneconomical to have the boiler running just to provide hot water. The basic type has a single element which projects downwards from the dome of the hot cylinder. Dual types have one short and one long element, controlled by a special dual switch, and the amount of hot water produced depends on which element is switched on. It is also possible to fit two short elements into the side of the tank, one high up and one low down, again to heat all or some of the cylinder.

An immersion heater needs its own independent power circuit, run from a 15-amp or 20-amp fuseway. It can be controlled by a time switch, and may be supplied with night-rate electricity via a white meter.

## What to do

There are several options for wiring up the heater, depending on where it is and whether one or two elements are fitted.

**Option 1** This is the most common option. Run the supply cable from the consumer unit to a 20-amp double-pole switch (Fig 129) near the hot cylinder. Use heat-resistant flex to link the switch to the heater terminals (Fig 132), via a time switch (Fig 130) if required.

**Option 2** This is used when the cylinder is in a bathroom, within reach of the bath. Here use a cord-operated ceiling switch instead. Run cable from it to a flex outlet (Fig 127) near the heater. Again, use heat-resistant flex to link this to the heater terminals. Add a time switch between flex outlet and heater if you wish.

**Options 3 and 4** These are used when a dual type or two separate heaters are installed. Run the supply cable to a special dual switch (Fig 131), and then run two lengths of flex to the heater terminals (Fig 133).

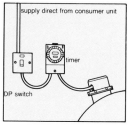

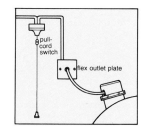

Fig 134 Heater wired via 20-amp DP switch and timer.

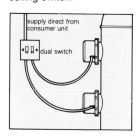

Fig 135 Heater wired via ceiling switch.

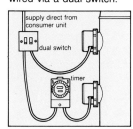

Fig 136 Two heaters wired via a dual switch.

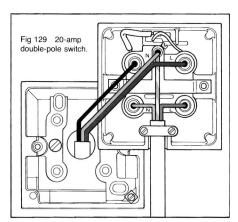

Fig 129 20-amp double-pole switch.

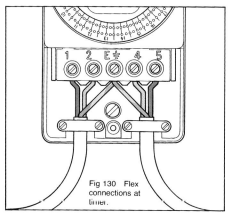

Fig 130 Flex connections at timer.

Fig 132 Wiring to single element immersion heater.

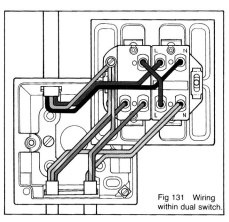

Fig 131 Wiring within dual switch.

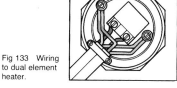

Fig 133 Wiring to dual element heater.

Fig 137 Two heaters, the lower one on a time switch.

# Wiring Cookers

Large free-standing cookers and built-in ovens and hobs consume large amounts of electricity, and so must be provided with their own dedicated circuits run from individual 30-amp or 45-amp fuseways in the consumer unit. The circuit cable is run to a point near the cooker, where it is connected to a suitably rated double-pole switch; from there, more cable is run to the cooker itself – see below.

The most important part of wiring for cookers lies in choosing the correct size of cable. In theory, a large cooker (or an oven-and-hob combination) could take as much as 50 amps if everything was switched on at once, and this would come close to blowing the house's main service fuse if other appliances were in use. In practice, however, you are very unlikely to have everything on at full blast, so a principle called *diversity* is used to make the calculations that determine the correct cable size. Here is how it works.

Start by finding out the wattage rating of the cooker – it is usually stamped on a plate at the back. Divide the rating by the mains voltage (240V) to get the maximum current demand. For a 12kW cooker, this would be 12,000 ÷ 240 = 50 amps. The diversity principle assumes the first 10 amps of demand is always needed, and assesses that on average you will use 30 per cent of the rest. So for our 12kW cooker, the current demand will be 100 per cent of 10 amps plus 30 per cent of 40 amps, which is 12 amps, giving a total current demand of 22 amps. If the circuit supplies a cooker control unit containing a socket outlet (*see* page 55), you add another 5 amps – that makes a total of 27 amps.

For a cooker of this size you should use 6mm² cable, with the cooker circuit protected by a 30- or 32-amp protective device and fitted with an isolating switch of the same rating. Use 10mm² cable and a 40- or 45-amp protective device for cookers rated at more than 12kW. *See* CHECK for further information on extra-long circuits.

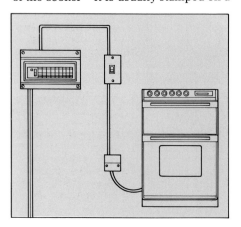

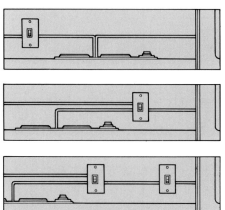

Fig 138  Wiring and switch arrangements for a free-standing cooker (*far left*) and for a separate oven and hob.

CHECK
- that you are using cable of the correct current rating for your cooker. For a cooker rated at up to 12kW, use 6mm² cable with a 30- or 32-amp protective device. Use 10mm² if the circuit is over 20m (66ft) long, up to a maximum circuit length of 30m (98ft).
  For cookers rated at more than 12kW, use 10mm² cable and a 40- or 45-amp protective device. Use 16mm² cable if the circuit will be more than 6.7m (22ft) long, up to a maximum circuit length of 10.2m (33ft).

## What to do

Option 1  Free-standing cookers are linked to the mains with a length of cable, not flex, because of the high current demand. Run the supply cable to a point no more than 2m (6ft 6in) from the cooker position, so it is within easy reach in case of emergencies. Mount a 30-amp or 45-amp double-pole switch at this point, connect in the circuit cable (Fig 140) and run another length of cable to a point on the wall at the back of the cooker recess. Fit a cooker connection unit here, and connect the supply cable to it (Fig 141). Finally connect a length of cable from the connection unit to the cooker terminals (Fig 142), making sure it is long enough to allow the cooker to be pulled out for cleaning without straining the cable.

Option 2  Built-in components can be controlled from one switch, but both components must be within 2m (6ft 6in) of the switch. You can run the cable from the switch to one component, then on to the next, or link each component directly to the switch. If the components are too far apart to meet the 2m requirement, fit two switches. Run the supply cable to the first switch, then loop it on to the second.

# Wiring Cookers

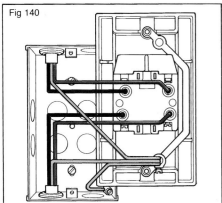

Fig 140

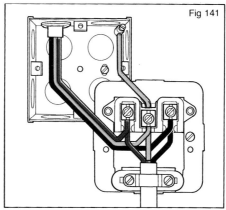

Fig 141

Fig 139 (*above*) Split-level appliances and free-standing cookers must have a control switch no more than 2m (6ft 6in) away.

Fig 140 How the incoming and outgoing circuit cables are connected into a double-pole cooker switch.

Fig 141 How the supply cable and the link to the cooker are wired up within a cooker connection unit.

Fig 142 How the cable connections are made to the terminals of a free-standing cooker.

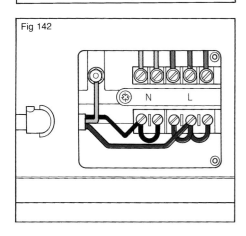

Fig 142

## Cooker Control Units

These units combine a double-pole switch with a standard 13-amp socket outlet, and are wired up in exactly the same way as a plain double-pole switch, with the circuit cable running on to a cooker connection unit behind the cooker. However, they have a major drawback: if they are mounted close to the cooker or hob position, the flex of any appliance plugged into the control unit socket could trail across a hotplate. Since this could be highly dangerous, their use is now being discouraged in favour of plain double-pole control switches.

# Wiring Instantaneous Showers

Electric showers are very popular because they are relatively easy to install, needing just mains-pressure cold water and an electricity supply – much simpler to arrange than the plumbing for a conventional shower. Like cookers, they are big current users and must also have their own dedicated power supply run from a separate fuseway in the consumer unit or fuse-box.

The circuit cable (Fig 143) runs first to a double-pole switch near the shower, and then to the shower heater where it is connected directly to the unit's terminal block. For showers with power ratings of up to 7kW and a current demand of about 29 amps, you should wire the circuit in 6mm² cable with a 30-amp cartridge fuse or miniature circuit breaker, or in 10mm² cable if a rewirable fuse is fitted. For showers with higher ratings, use 10mm² cable protected by a 45-amp fuse or circuit breaker, and make sure that the double-pole switch controlling it is also rated at 45 amps.

## What to do

Start by deciding where the shower is to be sited, so you can plan the route of the supply cable. Ideally this should enter the shower casing through a hole in the wall directly behind the unit. Connect the cable to the unit's terminals (Fig 145), and run it back to the shower switch.

For switching within the bathroom, it is best to use a ceiling-mounted switch with a cord pull, but a wall-mounted type can be used so long as it is at least 2m (6ft) from the bath or shower, or is sited outside the bathroom. Choose a switch with a neon indicator and a mechanical on/off flag, so you can see at a glance when the power supply to the shower is on.

At the control switch, connect the circuit cable to the terminals (Fig 144), and then run it back towards the consumer unit. For additional safety, include an in-line residual current device (see page 57). Finally, ensure that the shower supply pipework is cross-bonded to earth (Fig 146).

### What you need:
- shower unit
- 30-amp or 45-amp double-pole switch
- 6mm² or 10mm² two-core and earth cable, to suit shower rating
- earth sleeving
- 10mm² single-core earth cable for cross-bonding
- earthing clamp
- RCD (see page 57)
- electrical tools
- wood screws
- wall-plugs
- electric drill plus twist/masonry bits
- general building tools

### CHECK
- with your electricity board that your main service fuse can cope with the shower's current demand

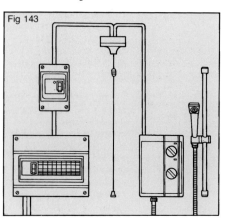

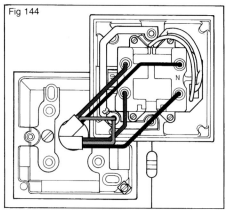

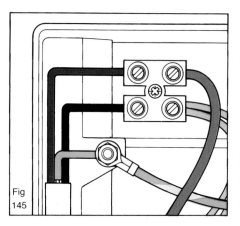

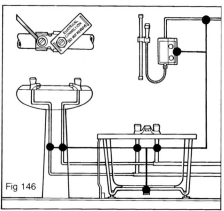

Fig 143   The circuit layout runs from a 30-amp or 45-amp fuseway via an in-line RCD to a double-pole switch, and then on to the heater terminals.

Fig 144   How the cables are connected within the ceiling switch.

Fig 145   How the supply cable is connected to the heater terminals. Note the single-core earth cable, which is cross-bonded to a clamp on the supply pipework.

Fig 146   All exposed metalwork in a bathroom must be cross-bonded to earth for safety.

# Using Residual Current Devices

Every electrical installation has a number of safety devices built in to guard against electrical faults and malfunctions. For example, circuit fuses or miniature circuit breakers (MCBs) are designed to cut off the power supply if the circuit is over loaded, or if a fault such as a short circuit occurs. However, they protect the circuitry more than the user, who is particularly at risk from electric shocks. Residual current devices (RCDs) provide precisely that protection, and are now a feature of all modern wiring installations.

RCDs, which used to be known by the even more cumbersome title of earth leakage circuit breakers (ELCBs), work by monitoring the flow of current in the circuit they are protecting. If they detect any imbalance between the flow in the live and neutral conductors, caused by current flowing to earth because of an insulation fault or because someone has touched something live, they cut off the current flow within a fraction of a second – fast enough to save a life.

RCDs are fitted within consumer units in modern wiring installation, providing protection for all the house's circuits. If you do not have one, you can get an electrician to install one alongside the consumer unit. Alternatively, you can fit one yourself to protect an individual circuit, by wiring it into the circuit cable at a convenient point.

The current wiring regulations require RCD protection for sockets used to power appliances for outdoor use – things like lawn-mowers and hedge-trimmers. So if you do not have whole-house RCD protection, you can at least provide this required level of protection by designating a particular socket for use with these appliances, and by replacing it with a special socket containing an RCD. These are the same size as a standard double socket, so fitting one involves a simple change-over.

The only snag you may hit is that some RCD sockets need mounting boxes 35mm deep, and your existing boxes may be only 25mm deep. Check this point before you buy the new RCD socket.

## What to do

**Option 1**   Fitting a socket outlet containing an RCD in place of an existing double socket is a very simple job. All you have to do is to turn off the power supply to the socket, unscrew its face-plate and disconnect the cable cores. You then reconnect the cores to the live, neutral and earth terminals of the RCD socket (Fig 148) and fit it back on its mounting box.

**Option 2**   An in-line RCD comes complete with its own enclosure, which you can mount anywhere that is convenient – on the same board as your consumer unit if there is room, or on a wall surface otherwise. To wire it into an existing circuit, turn off the mains supply and cut the circuit cable. Prepare the cable ends for connection, and wire the live and neutral cable cores to the RCD terminals (Fig 149) as directed by the maker's wiring instructions. The earth cores are wired to a connector bar. Then fit the cover and restore the power so you can test the operation of the RCD.

If you want an RCD to provide whole-house protection, call in a qualified electrician to specify the correct type for your installation and get him or her to fit it.

**What you need:**
- RCD socket outlet
- electrical screwdriver
- in-line RCD plus enclosure
- screwdriver
- wood screws
- wall-plugs
- electric drill plus twist/masonry bits
- side-cutters
- wire strippers
- earth sleeving

**CHECK**
- that the RCD's current and trip ratings are suitable for the purpose. Ask your supplier for advice

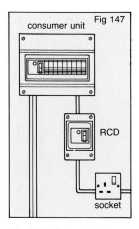

Fig 147

consumer unit

RCD

socket

Fig 147 (*above*) An in-line RCD provides protection for an individual circuit.

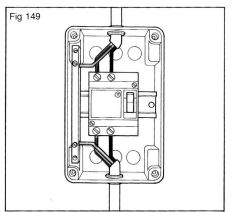

Fig 148

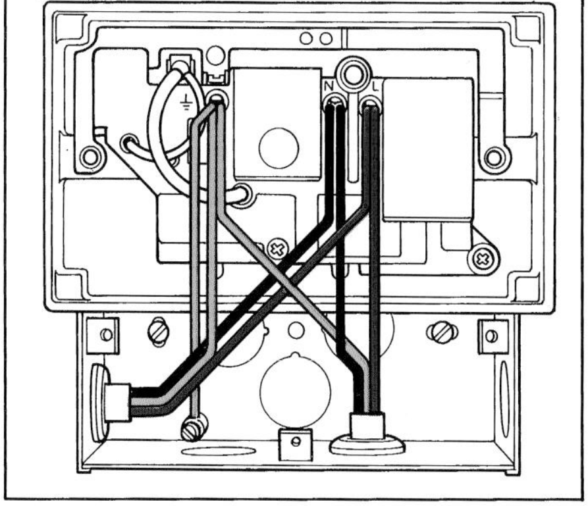

Fig 149

Fig 148   How the circuit cable cores are connected to the face-plate of an RCD socket outlet.

Fig 149   How the wiring connections are made within an in-line RCD.

# Adding a New Lighting Circuit

If you want to fit more lights around the house and your existing circuits are already supplying their maximum of eight lighting points, you will have to provide an additional lighting circuit. This should be run in 1mm² two-core and earth cable from a 5-amp fuseway – in your consumer unit, if it has any spare capacity, or else housed within a separate switch-fuse unit mounted alongside (see page 60 for details of how to install one).

Plan your new circuit carefully, so you can minimize the disruption that will inevitably be involved in running in the new circuit cables. This preliminary planning will also enable you to produce a detailed shopping list – light fittings, roses and lamp-holders, switches, junction and conduit boxes, connectors and so on. Estimate cable runs as best you can, and buy an extra 10 per cent to cope with unforeseen problems. If you need a new switch-fuse unit, buy this too. Note that a professional electrician must connect the new unit to the main supply.

## What to do

Remember that you can use a mixture of loop-in and junction-box wiring techniques to make the most economical use of cable.

Start by running the circuit cable from the consumer unit or switch-fuse unit to the various lighting positions, leaving plenty of slack at each point. Now install the lights themselves, fitting roses or conduit boxes as necessary, and connect the supply cables to them – either directly for loop-in wiring, or via strategically-placed junction boxes otherwise.

Next, wire in the switch cables. You will have to cut chases to new switch positions, but you may be able to feed cables down to existing switch positions if the old switch cables were run in conduit.

With the circuit wiring completed, make the main circuit connection to your consumer unit or switch-fuse unit. Turn the power off and connect the cable cores to the appropriate terminals.

**What you need:**
- 1mm² two-core and earth cable
- earth sleeving
- red PVC tape
- light fittings
- conduit boxes
- strip connectors
- ceiling roses
- pendant lamp-holders
- flex
- junction boxes
- light switches
- strip connectors
- electrical tools
- general tools

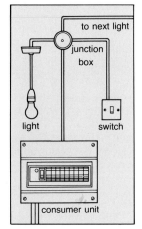

Fig 150 Junction-box wiring.

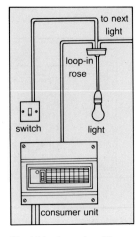

Fig 151 Loop-in wiring.

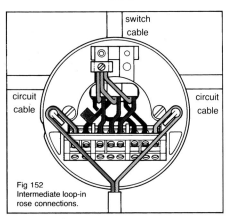

Fig 152 Intermediate loop-in rose connections.

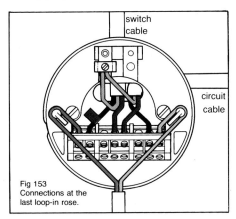

Fig 153 Connections at the last loop-in rose.

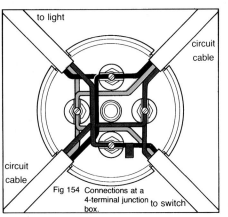

Fig 154 Connections at a 4-terminal junction box.

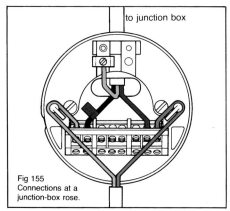

Fig 155 Connections at a junction-box rose.

# Adding a New Power Circuit

Power circuits, whether wired as ring mains or radials, can serve as many socket outlets as you wish, so you can generally satisfy your requirements by adding sockets to existing circuits (*see* pages 46 and 47). However, you are restricted in terms of the floor area of house which each power circuit can serve – 100 sq m (1,075 sq ft) for ring mains, 20 sq m (215 sq ft) for radials wired from a 20-amp fuseway and 50 sq m (540 sq ft) for radials on a 30-amp fuse. So if you plan to extend your house, you may have no alternative but to provide an additional power circuit.

As with a new lighting circuit, you need a spare fuseway, and if your consumer unit has no spare capacity you will have to add a new switch-fuse unit (*see* page 60). Then decide which sort of circuit you intend to install, according to the floor area it will have to serve. Ring mains, and radials serving between 20 sq m (215 sq ft) and 50 sq m (540 sq ft), need a 30-amp fuseway. Radials serving less than 20 sq m (215 sq ft) need a 20-amp one.

## What to do

Begin by deciding exactly where you want your new sockets to be, so you can work out how best to run the supply cables to them. If the area being served by the new circuit is remote from the consumer unit, it is generally better to wire it as a radial circuit, as this will avoid the necessity of completing the circuit with a long run back to the consumer unit.

Start at the consumer unit, leaving plenty of spare cable for the final connection to the mains, and run the cable to each socket position in turn. Remember that you can use spurs, either from sockets on the main circuit or from 30-amp junction boxes, to make the most economical use of cable.

Next, fit the mounting boxes, feed in the cables and connect the new sockets. Return to the consumer unit or switch-fuse unit, switch off the power at the mains and make the final connections to the new circuit's fuseway.

**What you need:**
- 2.5mm² two-core and earth cable for ring circuits and 20-amp radials
- 4mm² cable for 30-amp radials
- earth sleeving
- socket outlets
- mounting boxes
- 30-amp junction boxes
- fused connection units
- electrical tools
- general tools

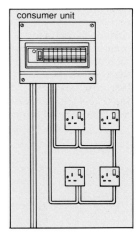

Fig 156   Ring circuit wiring.

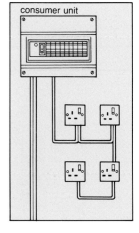

Fig 157   Radial circuit wiring.

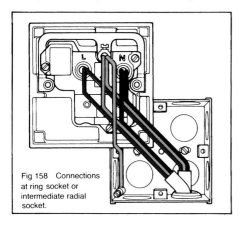

Fig 158   Connections at ring socket or intermediate radial socket.

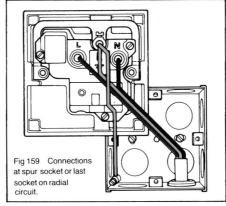

Fig 159   Connections at spur socket or last socket on radial circuit.

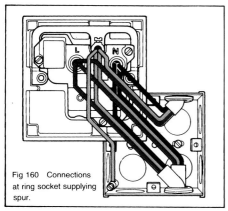

Fig 160   Connections at ring socket supplying spur.

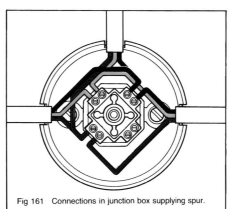

Fig 161   Connections in junction box supplying spur.

# Fitting a Switch-Fuse Unit

Unless you are lucky enough to have a modern consumer unit with some spare fuseways, adding extra circuits to your wiring system will mean installing a switch-fuse unit alongside your existing circuit board. Since you are going to some trouble to increase your system's capacity, it makes sense to fit a unit with enough extra fuseways to cope with any future extension work you might want to carry out. The only limit on the number of circuits you can have is the capacity of your system's main supply cable and service fuse to carry the extra current demand (see Check).

Connecting the power supply to the new switch-fuse unit is a job for a professional electrician; since this part of the system belongs to the electricity board, you are not allowed to tamper with it. The meter tails have to be disconnected from the consumer unit and taken to a distribution box. New tails (16 or 25mm², according to the load) are then run from there to both the existing and new units (Fig 162).

## What to do

Start by mounting the new switch-fuse unit next to the existing consumer unit, on the same backing board if possible or on a separate one nearby otherwise. Then connect up the new single-core live and neutral meter tails to its supply terminals (Fig 163) and run these back to the distribution box, also mounted on the unit's backing board.

Next, wire the new circuit cable(s) into the appropriate fuseway(s) in the switch-fuse unit (Fig 164), taking the live core(s) to the individual circuit fuse or MCB, and the neutral and earth core(s) to their respective terminals. Link the unit's earth terminal to the system's main earthing point. Then fit the cover, and call an electrician to reconnect the meter tails to the distribution box (Fig 165).

If you want the circuits supplied by the new switch-fuse unit to have RCD protection, pick a unit big enough to incorporate its own RCD alongside the circuit fuseways.

**What you need:**
- switch-fuse unit with MCBs (and integral RCD if required)
- new backing board (if needed)
- 16 or 25mm² single-core meter tails
- 10 or 16mm² single-core earth cable
- distribution box
- wood screws
- screwdriver
- side-cutters
- wire strippers

**CHECK**
- that your system can cope with the current demand of the new circuits. Your local electricity board will advise you

**TIP**
Even if your existing consumer unit contains rewirable fuses, make sure your new unit has MCBs – and ideally its own integral RCD.

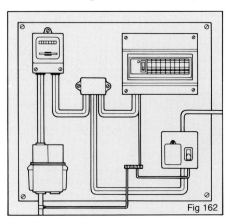

Fig 162

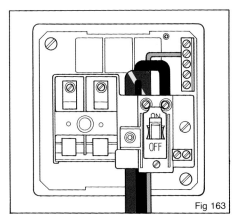

Fig 163

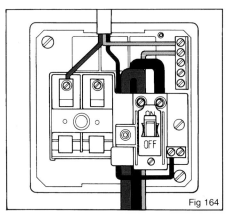

Fig 164

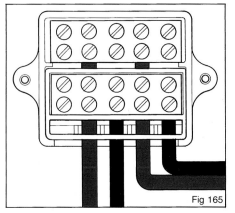

Fig 165

Fig 162   The new fuse-board layout, with the consumer unit and new switch-fuse unit both supplied by tails from a new distribution box.

Fig 163   How the meter tails are connected to the new switch-fuse unit.

Fig 164   How the new circuit cable is connected into the switch-fuse unit.

Fig 165   How the tails to the existing and new units are wired at the new distribution box.

# Wiring TV/FM Aerial Sockets

Unless you live in an area with very good television and radio reception, you will probably need a roof-top or loft aerial to guarantee adequate TV and FM signals. You can then link the aerial to socket outlets at convenient points around the house using coaxial cable.

A neat way to avoid having separate down-leads from your TV and FM aerials is to use diplexers. One is mounted in the loft, and leads from the two aerials are wired into it. Then a single down-lead is plugged into the diplexer and is run down to wherever you want to plug in your TV and hi-fi equipment. At this point a special combined TV/FM aerial socket is mounted over a single box, and the down-lead is connected into it. The socket has twin outlets, one for the TV and the other for the radio and comes in finishes to match other wiring accessories.

## What to do

Start by mounting your TV and FM aerials – either securely attached to a chimney stack or gable wall outside the house, or screwed to the roof timbers within the loft. Then connect lengths of coaxial cable to each aerial, and run them to the diplexer in the roof space. Connect the two leads to the diplexer. Fit a coaxial plug to the common down-lead as mentioned earlier, plug it in and run the lead down to where the socket is to be installed. It can be run inside the house, just like any other circuit cable, or else taken down the outside wall and through a hole into the house.

Fit the mounting box, run in the down-lead and connect it to the socket terminals (Fig 167). Secure the socket to its mounting box. Then plug in the leads to your TV and radio receiver to complete the job.

What you need:
- TV and FM aerials
- mounting brackets
- diplexer
- coaxial cable
- TV/FM aerial socket
- mounting box
- electrical tools

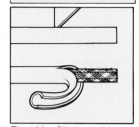

Fig 169   Slit the cable sheath, peel back and cut away.

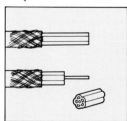

Fig 170   Roll back the screen and trim the insulation.

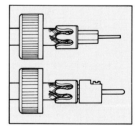

Fig 171   Fit the cap, screen grip and pin moulding.

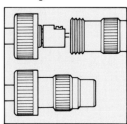

Fig 172   Slip on the pin body and screw down the cap.

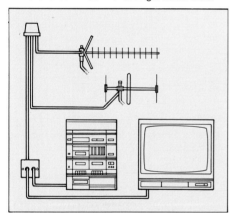

Fig 166   Diplexers allow aerials and receivers to be linked with a single down-lead.

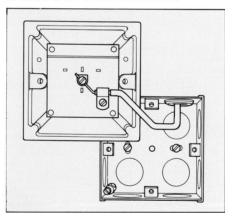

Fig 167   The down-lead connection to the terminals of the TV/FM socket outlet.

Fig 168   Unless you live in a good reception area, you will probably need an aerial on your roof-top to guarantee good TV and FM signals.

# Wiring Telephone Sockets

Householders are now allowed to wire up extension telephone sockets themselves, and the skills involved are very similar to those used in other electrical wiring work. Kits are available containing everything needed for the job – extension sockets, connectors, six-core cable, clips and even a special tool for connecting the flimsy cable cores to the socket terminals. However, before you can install extension sockets, you must have a modern, square master socket fitted by your telecommunications operator. This is one job you are not allowed to do yourself.

You can add as many extension sockets as you need, but your system won't work if you plug more than four phones at a time into it. The sockets can be wired from a junction box or in series (Fig 174); choose whichever layout suits your needs and makes the most economical use of cable. You can buy single and double outlets to match most wiring accessory ranges.

## What to do

Start by deciding where you want to site your extension sockets, then mount the sockets in place (flush- and surface-mounting boxes are available). Start work at the master socket, leaving enough slack for the plug-in converter to be connected. It's simplest to surface-mount the cable, clipping it to the tops of skirting boards and running it round door architraves, but you can chase it into walls and run it under floors if you want to.

Run the cable to each socket position in turn, leaving plenty of slack at each point. To prepare it for connection, slit the sheath lengthwise and locate the fine draw-string inside. Draw it out, wrap it round your fingers and pull it along to split the sheath. Then cut off the waste. Cut back the cores and press them into the socket terminals, using the connector tool provided. Note that the colour-coded cores are always connected like to like.

**What you need:**
- extension kit *or* extension sockets
- mounting boxes
- joint boxes
- plug-in converter and attached cable
- extra cable
- cable clips
- connector tool
- wood screws
- wall-plugs
- screwdriver
- handyman's knife
- general tools

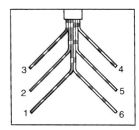

Fig 173   Follow this diagram to help you connect the cores to their correct terminals.

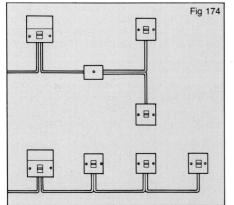

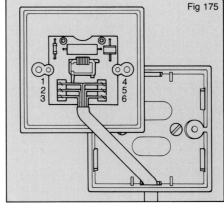

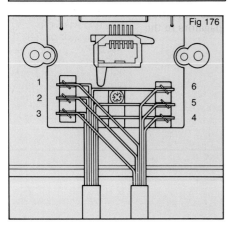

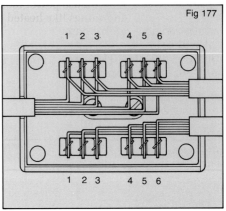

Fig 174   You can wire your sockets as branches from a joint box (top) or in series (bottom).

Fig 175   At a single extension socket, connect the cores as shown – coloured cores with white bands (1,2,3) to the left, white cores with coloured bands (4,5,6) to the right.

Fig 176   At intermediate sockets, add the second set of cores, connecting like to like.

Fig 177   At joint boxes, connect the cores as shown, again linking like to like.

# OUTDOOR JOBS

Electricity can be just as valuable a servant out of doors as it is indoors. Lighting lets you (and your visitors) find the way to the front door on dark evenings, lights up the patio for summer parties, shows off the garden and also helps to keep would-be burglars away. An outdoor power supply enables you to use the growing range of powered garden tools safely and conveniently, and taking power to out-buildings such as workshops and greenhouses will greatly enhance your enjoyment of your hobbies. All these jobs are covered in this section.

However, the great outdoors can be a dangerous place as far as electricity is concerned. Not only must everything be carefully protected against the elements to stop water getting into the wiring; you need protecting too against the risks of electric shock.

## What to Tackle First

As with indoor jobs, you can divide your outside wiring work into the compulsory and the voluntary. Some things will greatly improve your quality of life, others will simply be good to have. Here are some ideas for you to consider.

**Lighting for access**   Start by installing a light next to the front door, which lets you see and identify callers as well as helping to light the way from the front gate. If it also lights up the house number or name, so much the better.

Add lights along the front path if necessary, especially if it contains any steps. Position the lights to shine on the risers of each step.

Consider fitting a passive infra-red detector to control the lights. It will detect anyone approaching the houses and switch the lights on for a pre-set period, before turning them off again.

**Lighting for pleasure**   It's a simple job to add lights on the back wall of the house to light up the patio, and not very complicated to wire up more lights remote from the house, especially if you use low-voltage circuits which are quick to install

and completely safe to use.

**Power supplies in the garden**   Having one or more outdoor socket outlets, either on the house walls or at various points around the garden, makes it simpler and safer to use powered garden tools since you do not need long extension leads run out across the garden from a socket inside the house.

**Power to outbuildings**   If you use a shed as an outdoor workshop, or have a greenhouse and raise your own plants from seed, a power supply can be a great boon. It can extend the amount of time you can spend in the building by allowing you to have light and heating, and also lets you use power tools in the workshop and things like heated propagators in the greenhouse.

## Safety out of Doors

If you plan to use electricity out of doors, take great care to ensure that the wiring is carried out to the highest standards and that only weatherproof fittings are used. Most importantly, make sure that circuits and socket outlets all have RCD protection (see page 57).

Fig 178 (*above*)   Fitting an outside light is just one of the many wiring jobs you can carry out out of doors.

**NEW REGULATIONS**
From January 2005, all new outdoor wiring work must meet the requirements of the new Part P of the Building Regulations. All the jobs in this section **must** be notified in advance to your local authority Building Control Department. You also have to pay a building control fee to have the work inspected and tested by an electrician when you have completed it. See page 6 for full details.

# Fitting a Light on the House Wall

It is a great boon having lights out of doors, especially to light the way to your front door, or to illuminate the patio on summer evenings. There is nothing worse for after-dark visitors than having to negotiate an unfamiliar and unlit path. An outside light, however, not only lets them see where they are going, it also makes it far easier for them to find your house in the first place. Lighting up the patio is more for your pleasure than anyone else's, and can greatly extend the use you get from it, whether you are sitting alone with a book and a drink or entertaining friends at a barbecue.

The other advantage of outside lighting is that it helps to deter would-be intruders. By keeping your house and its surroundings well lit, you get the benefit of greatly improved home security at virtually no extra cost – you needed the lighting anyway. You can wire the new lights up in one of two ways.

Remember that all the fittings you use must be suitable for outdoor installation.

## What to do

**Option 1**  The first option is to wire your new light(s) as an extension of an existing lighting circuit (Fig 179), so long as this will not overload it; remember that each circuit should not supply more than eight lights (see pages 38 and 39 for more details).

Assuming that you can extend a circuit, start by locating the main circuit cable at a point as close as possible to the point where you want to install the new outside light. Then turn off the power (see page 34), cut the circuit cable and connect the cut ends to a four-terminal junction box (Fig 180), screwed to the side of an underfloor joist.

From there run a spur cable to the new light position, and another cable to where the new light switch is to be fitted (Fig 181). Take the light cable through a hole drilled in the house wall and lined with some 16mm diameter PVC conduit. Then connect the cable to the light, either directly (Fig 182) or via strip connectors (Figs 186 and 187).

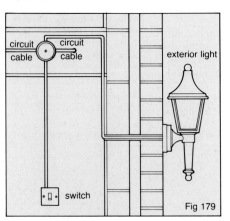

Fig 179

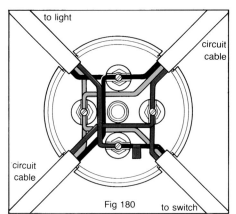

Fig 180

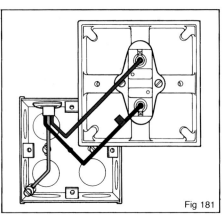

Fig 181

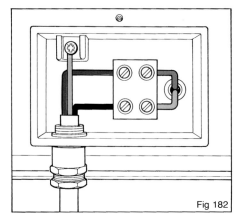

Fig 182

Fig 179  How to extend an existing lighting circuit to supply a new outside light.

Fig 180  How the cables are wired within the 4-terminal junction box.

Fig 181  The connections within the new light switch. Note the red tape flag on the neutral switch cable core.

Fig 182  How the spur cable to the light is connected if the fitting has its own terminal block.

# Fitting a Light on the House Wall

**Option 2** The alternative way of supplying power to outside lights is to run a fused spur from a nearby power circuit – the only option if your lighting circuits are already serving their maximum of eight lights. Locate a convenient circuit (the upstairs power circuit is generally the easiest to extend), and either run your spur cable from a socket on the circuit (Fig 184) or use a three-terminal junction box to connect it directly to the power circuit itself (*see* pages 46 and 47 for more details).

Run the spur in 2.5mm$^2$ cable as far as the fused connection unit (Fig 185), and fit it with a 5-amp fuse. Then run 1mm$^2$ cable on from the FCU to a four-terminal junction box, and on to the light and switch. The wiring at the junction box, light and switch are identical to that for extending a lighting circuit (Figs 180 to 182).

Alternatively, you may be able to site a switched FCU so it can do double duty as the new light's on/off switch.

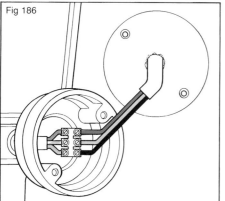

Fig 184

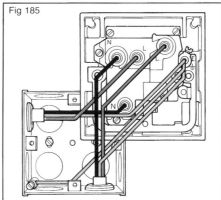

Fig 185

Fig 183 (*above*) Outside lights are a boon for visitors and a deterrent to would-be intruders.

Fig 184 How to extend an existing power circuit to supply a new light.

Fig 185 How the incoming and outgoing cables are connected at the FCU.

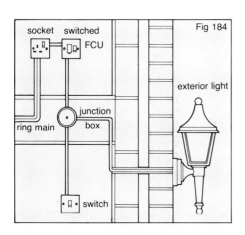

Fig 186

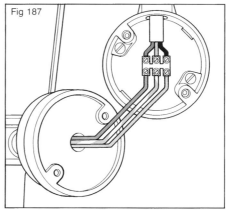

Fig 187

Fig 186 Some light fittings have flex tails and a recessed base. Connect the circuit cable as shown, and mount the light on the wall.

Fig 187 Other fittings have a flat base, so you need a conduit box recessed into the house wall to contain the wiring connections.

# Fitting other Outside Lights

If your new lights will be remote from the house, as opposed to being mounted on the house walls, you must wire them on a separate circuit; you are not allowed to extend your house wiring circuits beyond the house itself. You can use a spare fuseway in your consumer unit for the new circuit, or else add a new switch-fuse unit (see page 60 for more details). There is of course nothing to stop you from wiring lights on the house walls via their own circuit if you prefer.

You can mount exterior light fittings on walls, or on individual posts (but never on posts that also support a fence – a gale could blow the fence run down and sever the supply cable). Some fittings come with an integral ground spike, and are intended to be positioned in lawns or flower beds to illuminate individual garden features. These can be connected permanently to the mains if you wish, or can be plugged into outdoor sockets at various points around the garden.

## What to do

**Option 1** Mains-voltage wiring. Start by deciding where you want to site the new lights, and what type of fitting you intend to use, so you can plan the cable runs to them. You can run cables along exterior walls, but they must not be fixed to fences. Otherwise, they should be run underground in PVC conduit, set in a trench 450mm (18in) deep where it crosses cultivated areas such as lawns and flower beds to protect it from accidental damage from garden tools.

Where lights are to be set on walls, pillars or post, the cable is normally run directly into the fitting via a weatherproof rubber seal. The conduit should run right up to the fitting itself. Movable lights on ground spikes are normally supplied with flex tails, and if they are going to be connected permanently to the mains you need to mount a conduit box on a post near the lamp position, so that the flex can be

### What you need:

*Option 1*
- new light fittings
- 1.5mm$^2$ two-core and earth cable
- earth sleeving
- impact-resistant PVC conduit plus fittings
- conduit boxes
- silicone mastic for sealing boxes and cable entries
- 4-terminal junction box
- one-gang one-way switch
- in-line RCD
- electrical tools
- general tools

*Option 2*
- light fittings
- low-voltage cable
- transformer
- 13-amp plug with 3-amp fuse

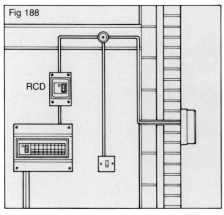

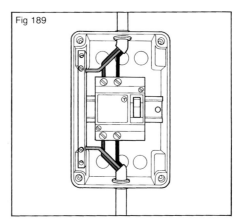

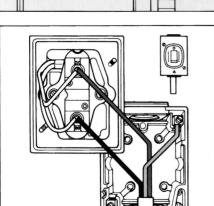

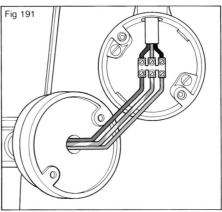

Fig 188   How a separate circuit is run to supply outside lights via an in-line RCD.

Fig 189   How the circuit cable is connected within the RCD.

Fig 190   How the cable is connected within an outdoor switch.

Fig 191   With some types of outdoor light, the flex tails must be connected to the supply cable within a conduit box. Seal the joint between fitting and box with silicone mastic.

# Fitting other Outside Lights

connected to the circuit cable within a sealed enclosure (Fig 191). Otherwise the flex tails should be connected to a weatherproof plug so the lamp can be plugged into an outdoor socket outlet when you want the light on (see page 68).

Run the cable in conduit, buried where necessary, from the light position(s) back towards the house, and pass it through the house wall. Make sure the joints throughout the conduit run are sealed by solvent-welding, and line the hole in the house wall with conduit as well to prevent the cable sheath from chafing where it passes through the wall.

Within the house, run the circuit cable to a conveniently sited four-terminal junction box, wired up as shown in Fig 180 on page 64. Run cable from here to the switch controlling the lights; this can be sited indoors or outdoors, using a weatherproof switch (Fig 190) in the latter case. Complete the circuit by running cable from the junction box via an in-line RCD (Fig 189) back to the circuit fuseway at the consumer unit or switch-fuse unit if no spare fuseways are available.

**Option 2** An alternative to mains-voltage outdoor wiring is to use special low-voltage fittings run from a transformer. This has several advantages over a mains-voltage installation. The low-voltage cable can be run on the surface, so installation is quicker. The fittings consume a negligible amount of electricity. Above all, the system is totally safe if the cable is damaged or a fault develops.

All you do to install such a system is run out the cable, connect the fittings to it (Fig 193 and 194) and link it to the transformer (Fig 195).

Fig 192 Low-voltage lights are quick and easy to install, cheap to run and provide excellent illumination.

**CHECK**
- that all fittings used out of doors are labelled as suitable for exterior use

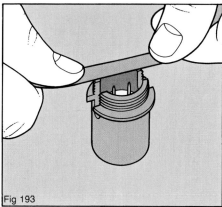

Fig 193

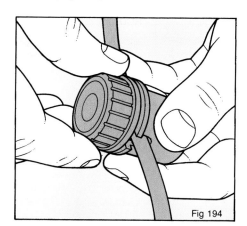

Fig 194

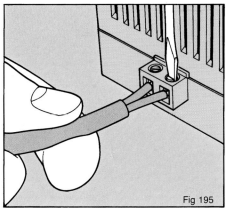

Fig 195

Fig 193 Most low-voltage fittings have pin connectors. To connect the supply cable, press it onto the pins.

Fig 194 Screw on the weatherproof cap to complete the connection.

Fig 195 Connect the low-voltage cable to the transformer, which must be sited indoors or in a weatherproof enclosure, and plug it into the mains supply.

# Fitting Outdoor Sockets

If you use a lot of garden power tools – lawn-mowers, hedge-trimmers and the like – you will appreciate the convenience of having one or more socket outlets out of doors. This not only avoids the nuisance and safety risks of trailing flexes from the house through doors and windows; it also ensures maximum safety for anyone using the tools should a fault develop, because any socket that will supply tools or equipment used out of doors must have RCD protection to satisfy the requirements of the wiring regulations.

Sockets on the back wall of the house or down the garden are also useful for other purposes. For example, it means you can safely take other domestic appliances out of doors – the toaster and the coffee percolator for an alfresco breakfast on the patio, for example, or the vacuum cleaner to give the car a spring-clean. You can also use outdoor sockets to plug in movable outside light fittings at various points around the garden.

## What to do

Start by deciding where you want to site the outdoor socket(s). If you just want one on the wall of the house, you can run the power supply to it as a spur from an indoor power circuit (Fig 196). The new socket must have RCD protection, and you can provide this either by wiring an in-line RCD into the spur cable as shown, or by fitting a special outdoor socket containing its own integral RCD.

If you intend to install sockets remote from the house, you must provide a separate RCD-protected circuit run from a spare 30-amp fuseway in the consumer unit, or else from a separate switch-fuse unit (Fig 197). Run the circuit cable in PVC conduit buried at least 450mm (18in) beneath open ground, and position sockets on walls or on individual posts set in the ground. Take the conduit right up to the socket position, and ensure that a weatherproof cable entry is used.

### What you need:
- weatherproof socket(s)
- impact-resistant PVC conduit and fittings
- 2.5mm² two-core and earth cable
- earth sleeving
- in-line RCD
- electrical tools
- general tools
- extension lead
- plug-in RCD adaptor

### CHECK
- that any socket used out of doors is marked as suitable for exterior use, and is protected by an RCD
- that all conduit joints are solvent-welded
- that all cable entries are fitted with weatherproof seals

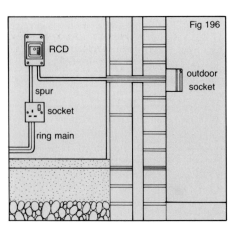

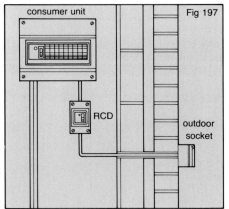

## Temporary Power Supplies

If you need a temporary power supply out doors and don't have an outdoor socket outlet, you will have to run an extension lead out from the house. If possible, use a proper extension reel with easily visible orange flex, and be sure to uncoil it fully; otherwise it may overheat within the drum, melting the insulation and causing a short circuit. Never use an extension lead out of doors in wet conditions. Most important of all, either plug the lead into a socket with RCD protection, or else use a plug-in RCD adaptor at the house socket if you do not have an RCD socket.

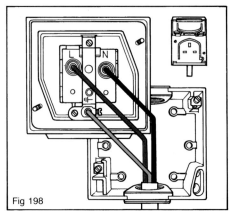

Fig 196   A socket on the house wall can be wired as a spur from an indoor power circuit, but it must have RCD protection.

Fig 197   Sockets remote from the house must be wired on a separate circuit run via an RCD from a spare fuseway in the consumer unit.

Fig 198   How the circuit cable is connected within an outdoor socket outlet.

# Taking Power to Outbuildings

A power supply to an outbuilding must be protected from the elements and should incorporate various safety requirements. The supply must be run as a separate circuit from your consumer unit or fuse box, and should be protected by a residual current device (RCD).

The cable run to the outbuilding can be taken overhead or underground, whichever is the more convenient. Overhead cables need no support for short spans, but must be carried on a catenary wire over longer distances. If taken underground, the cable should be buried at least 450mm (18in) down, and should be run in PVC conduit.

At the house end of the circuit, the cable originates either at a spare 30-amp fuseway in the consumer unit, or at a separate switch-fuse unit if there is no spare fuseway. It then runs via an in-line RCD out of the house towards the outbuilding by its chosen route. At the outbuilding end of the circuit it runs to another switch-fuse unit which allows the supply to be isolated at this end of the circuit. This switch-fuse unit can supply a 20-amp radial circuit to socket outlets, with a fused spur supplying the lights, or else it can supply a separate lighting circuit via a 5-amp sub-circuit fuse housed within the outbuilding's switch-fuse unit.

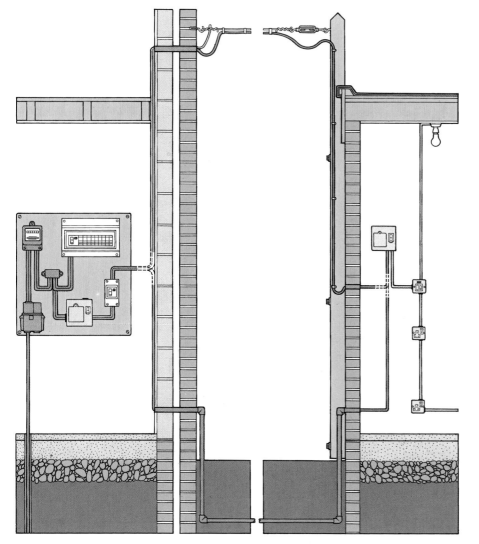

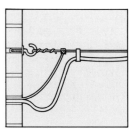

Fig 199   At the house end, the catenary is linked to the main earthing point.

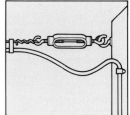

Fig 200   At the outbuilding, a tensioner keeps the wire taut. Drip loops allow rain to run off.

Fig 201   Circuits to outbuildings can be run overhead or underground, and must originate at a separate circuit fuse within the house. An in-line RCD must be fitted to protect the circuit.

69

# Taking Power to Outbuildings

## What to do

**Option 1**   The advantage of overhead wiring is that it is fairly quick and easy to install. However, it does not look particularly attractive, and is vulnerable to damage.

The cable run from inside the house to inside the outbuilding must be unbroken. It can be run in ordinary PVC-sheathed cable; the 2.5mm² size is adequate for outbuildings close to the house, but the 4mm² size should be used for longer runs. The cable needs no support over spans of less than 3.5m (11½ft) but for longer spans you must suspend it from a catenary wire – a length of galvanized steel wire run between secure fixings.

Since the cable must be at least 3.5m (11½ft) above ground level (and 5.2m, or 17ft, where it crosses drives or other vehicular access) you will have to erect a stout post on the outbuilding wall to carry that end of the wire, and make the exit point from the house the required height above ground.

Because the wire is under tension, you need secure fixings at each end. At the house end, use an expansion anchor fitted with a screw eye, and fit an eye bolt and nut to attach the other end to a timber post. Then attach the wire to the bolt at the house end of the run, take it to the tensioner at the outbuilding end and pull it as tight as you can. Next, tighten the tensioner to pull the wire taut.

You can now loop the cable into place, attaching it to the wire with proprietary ties. Then attach a length of single-core earth wire to the end of the catenary wire, ready for connection to the house earthing point at the consumer unit.

Feed the circuit cable and earth loop into the house through a hole in the house wall, ready for connection at the end of the job. At the outbuilding end, take the cable in and connect it to the terminals of the switch-fuse unit (Fig 205). Then wire up the circuits within the outbuilding. You can either have a single 20-amp circuit supplying socket outlets and a fused connection unit, which in turn feeds a spur to the light (Fig 202), or you can wire a separate 20-amp power circuit and a 5-amp lighting circuit (Fig 203).

**Option 2**   Running cable underground has the major advantage in that the run is protected from accidental damage. The installation is, however, likely to involve more work, especially if paths have to be dug up.

Start by planning the line of the cable run carefully, so that you do not have to do any unnecessary excavation work. Then drill the cable exit point from the house and the entry point to the outbuilding, and dig a trench between them. Make it 450mm (18in) deep if below open ground, shallower if the run will be paved over. Lay out sufficient cable to reach from the outbuilding switch-fuse unit to the consumer unit in the house, and make the main connection within the outbuilding. Make sure you clip the cable run from the switch-fuse unit securely to the walls of the outbuilding.

You can now start threading fittings and lengths of conduit on to the cable to make up the underground section of the run, solvent-welding each joint as you work. When the run is complete back to the house wall, back-fill the trench and complete the wiring within the outbuilding, as for overhead wiring.

**What you need:**

*Option 1*
- main circuit cable (2.5mm² or 4mm² according to length of cable run)
- galvanized catenary wire
- cable ties
- earth clamp
- earth sleeving
- eyebold fixings
- tensioner
- support post
- electrical tools
- general tools

*Option 2*
- main circuit cable
- earth sleeving
- impact-resistant PVC conduit and fittings
- solvent-weld cement
- electrical tools
- general tools

*For both options –
outbuilding circuits*
- switch-fuse unit
- metal-clad sockets
- metal-clad FCU *or* 5-amp junction box
- metal-clad switch
- mounting boxes
- 2.5mm² cable for power circuit
- 1mm² cable for lighting circuit
- cable clips
- light fitting

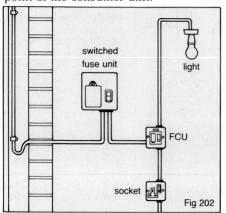

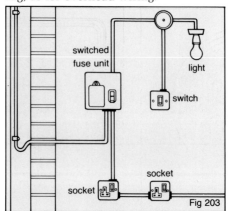

Fig 202   In the outbuilding you can run the light as a spur off the circuit to the sockets.

Fig 203   Alternatively, run a separate 20-amp circuit to the sockets and a 5-amp circuit to the light.

# Taking Power to Outbuildings

Finally, complete the installation by making the mains connections to the consumer unit or switch-fuse unit at the house end of the new circuit, and connect the earth link from the catenary wire to the house system's main earthing point.

Note that the mains supply to a new switch-fuse unit must be connected by a professional electrician.

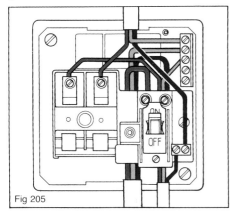

Fig 205

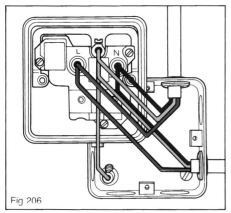

Fig 206

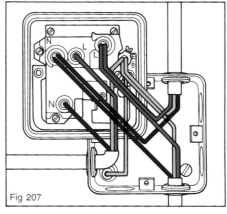

Fig 207

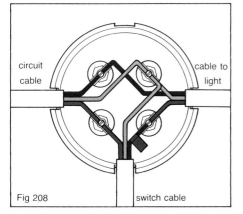

circuit cable

cable to light

Fig 208

switch cable

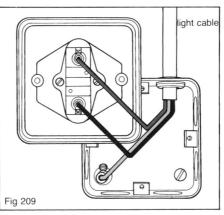

light cable

Fig 209

Fig 204 (*above*) A power supply is a must for garages, garden workshops and greenhouses.

Fig 205 How to wire the circuit cable(s) into the switch-fuse unit.

Fig 206 How to wire sockets in series on a 20-amp radial power circuit.

Fig 207 How to wire the lighting spur into the fused connection unit.

Fig 208 How to wire the junction box on a separate light circuit.

Fig 209 How to wire the light switch on a separate circuit.

# REPAIR JOBS

Electricity is clean, efficient, and almost universally available. It does its job so well that we take it completely for granted – until something goes wrong. The trouble may be caused by a fault developing inside an individual appliance, in the plug or flex connecting it to the mains, or with the mains supply itself. It may be the result of an accident, such as drilling through a hidden cable or tripping over a flex, or of simple wear and tear causing something to malfunction.

Whatever the problem, it is well worth trying to track it down and put it right yourself. Of course, not everything is within the scope of the do-it-yourself electrician, but many repairs are simple to carry out and doing them yourself will certainly save you money. Fixing things promptly when they go wrong is also good practice from the safety point of view.

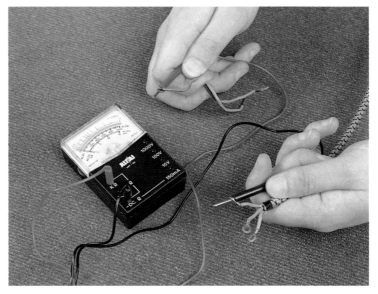

## What to tackle

The golden rule when tracing any electrical fault is to be methodical – and careful. By following the check-lists on pages 73 and 74, you will usually be able to work out what is causing a particular fault, even if you are unable to put it right yourself. That can at least put your mind at rest and also let you get everything else working again once you've isolated what is causing the problem. If it does enable you to repair the problem yourself, you will have saved yourself money, time spent in waiting for a repair man to come, and also face – it's surprising how many experts report being called out to fix faults that involved nothing more than remaking a connection or fitting a replacement fuse, to the customer's great embarrassment.

However, don't be tempted to try things beyond your capabilities. This point applies particularly to repairs to electrical appliances. Many of them are a great deal more complicated than they used to be, and repairs are generally best left to a specialist repair agent. Indeed, often this is imperative if the terms of the appliance guarantee are not to be rendered invalid by unauthorized tampering. The repairs

dealt with in this section are of an essentially general nature; leave anything not covered here to an expert.

Whenever you are investigating an electrical fault, never assume that something is safe to touch just because it is not working. Always unplug an appliance from the mains before opening it up, and turn off the power supply at the mains before exposing any part of your circuit wiring. As an additional safeguard, always use a mains tester to check that the power is off, and work using tools with insulated handles wherever possible.

## Who Owns What

Remember that the incoming mains supply cable, the service head and the electricity meter are all the property of your electricity distribution company, and that it is an offence to tamper with them. You are also not allowed to disconnect the meter tails leading to your consumer unit or fuse-box. This job must always be carried out by a qualified electrician.

The distribution company has the right not to connect a supply to a domestic wiring system that is unsafe, so keeping yours in good order is vitally important.

Fig 210 (*above*) Many electrical faults are relatively straightforward to track down, and are often surprisingly easy to put right yourself.

### NEW REGULATIONS
None of the electrical repair jobs described in this section need be notified to your local authority Building Control Department. According to the requirements of Part P, you should still have the work inspected and tested by an electrician, although Building Control Departments accept that in practice this is unenforceable.

# Tracing Electrical Faults

## If a Light doesn't Work

1.   Start by suspecting that the light bulb has failed. Turn off the light switch, remove the old bulb and fit a replacement of the appropriate type and wattage. Turn the switch back on. If it still does not work, go to Step 2.

2.   Your next move is to check whether the lighting circuit fuse or MCB has cut off the supply to the circuit. This may have happened because of a short circuit somewhere (common causes include a loose connection within a pendant lamp-holder caused by breezes making the light sway about, and overheating inside enclosed fittings). If the fuse has blown, turn off the main switch and remove the fuse-holder, fit new wire or a replacement cartridge fuse of the correct rating (5-amp — see page 75) and replace the fuse-holder. Restore the power and turn on the light. If the fuse blows again, go to Step 3.

3.   With the power off, remove the ceiling rose and lamp-holder covers of pendant lights and look for any loose connections or broken cores. Within enclosed fittings, check that insulation is intact on flex cores, and repair any damaged insulation with PVC insulating tape. Remake any loose connections that you find in roses or lamp-holders, check that flex cores are hooked over their anchorages in the rose and lamp-holder to prevent stress on the connections, and replace the covers. Restore the power and try the light again. If the fuse blows again or the light still does not work, go to Step 4.

4.   With the power off again, disconnect the pendant flex from the rose and lamp-holder terminals, and use a continuity tester to check that each core is unbroken. If any is, replace the flex (see page 76) and restore the power. If the fuse blows again, go to Step 5.

5.   Repeat Steps 3 and 4 for other lights on the circuit. If the fuse still blows, read about fault-finding on whole circuits on page 74.

## If an Appliance doesn't Work

1.   So long as the appliance has simply stopped working and there is no smell or sign of burning, plug another appliance or a portable light into the socket. If this works, the fault is with the original appliance; go to Step 2. If it does not, read about fault-finding on whole circuits on page 74.

2.   Unplug the appliance from the mains, undo its plug and check that the connections to the terminals are secure. Loose connections cause sparking and charring within the plug. Re-make them as necessary. Replace the plug top and plug the appliance in. If it still fails to work, go to Step 3.

3.   Unplug the appliance and reopen the plug. Remove the plug fuse and test it with a continuity tester or a metal torch (see Tip 1). Replace it, if it is faulty, with another of the appropriate rating for the appliance (a 3-amp fuse for appliances rated at less than 720 watts, a 13-amp one otherwise). Replace the plug top. If the appliance still fails to work, go to Step 4.

4.   Unplug the appliance again and open its outer casing if you can. Locate the terminal block to which the flex is connected, and check the connections. Remake them if they are loose, replace the appliance casing and plug it in again. If it still fails to work, go to Step 5.

5.   Again unplug the appliance and open the casing. Use a continuity tester to check the continuity of each flex core between the terminal block and the plug, and replace the flex (see page 77) if any core is faulty. Test the appliance again. If it still does not work, take it to a service engineer for testing and repair.

Note   If every socket on a power circuit is dead, read about fault-finding on whole circuits on page 74.

**What you need:**
- fuse wire or cartridge circuit fuse
- electrical screwdriver
- side-cutters
- wire strippers
- PVC insulating tape
- continuity tester
- replacement flex
- replacement plug fuse

### CHECK 1
- whether plug fuses have blown using a torch with a metal body if you do not have a continuity tester. Unscrew the end of the torch and hold the fuse so one end is touching the torch case and the other the end of the battery. Switch the torch on; a sound fuse will light it, but a blown one will not
- whether the fuse has blown in a dimmer switch if its light does not work. Replace it if it has.

### TIP 1
When fitting replacement bulbs, check to see if there is a label on the shade or fitting indicating the maximum recommended lamp wattage. If there is, do not exceed it.

   If a bulb is shattered and still in its lamp-holder, turn the power to the circuit off at the mains and use long-nosed pliers to grip and remove the cap from its socket.

# Tracing Electrical Faults

## If a Whole Circuit is Dead

1.  Switch off all the lights on the circuit, or disconnect all appliances plugged into it. Then check the appropriate circuit fuse-way to see whether the fuse has blown or the MCB has tripped off. If a fuse has blown, turn off the power and replace the fuse with new wire or a cartridge of the appropriate rating for the circuit (see page 75). Restore the power and go to Step 2. If the fuse blows immediately, go to Step 3. If an MCB has tripped off, reset it to 'on' and go to Step 2. If it cannot be reset, indicating that the fault is still present, go to Step 3.
2.  Switch on lights or plug in appliances one by one. If a particular light or appliance blows the circuit fuse or trips the MCB, isolate it for checking as described in fault-finding when light bulbs or appliances don't work on page 73. An individual appliance may not be faulty in itself, but may simply be overloading the circuit; take care not to connect too many high-rated appliances to any one circuit. Restore the power again. If the fuse blows once more or the MCB cannot be reset, go to Step 3.
3.  With the power to the circuit turned off, open up switches, socket outlets and other accessory face-plates so you can check for loose connections or short circuits caused by overheating or damaged insulation. Remake loose connections and repair damaged insulation using PVC insulating tape. Replace the face-plates and restore the power. If the fuse blows or the MCB trips off again, go to Step 4.
4.  The fault is somewhere else on the circuit wiring. If you have drilled through a cable or driven a nail into one, go to Step 5. If there is no other obvious explanation for the fault, call in a professional electrician to test the circuit for you.
5.  If the cable you have pierced is buried in a wall, you will have to replace the damaged section. If it is beneath a floorboard, lift the board, cut the cable at the point where the damage occurred and reconnect the ends using a junction box. See page 79 for more details.

## If the Whole System is Dead

1.  Where you appear to have a total power failure – in other words, none of your circuits have any power – and your system is protected by a residual current device (RCD), start by checking whether it has tripped off. If it has, reset it to 'on'. If you cannot reset it, this indicates that the fault which caused it to trip in the first place is still present. Run through the points under the other fault-finding sections to locate a possible fault, or call in a professional electrician to check the system over.
2.  If you do not have an RCD, check with neighbours to see whether there is a local power cut. This may not affect every house in the road if it is a single-phase supply fault. With the three-phase supply system used to distribute electricity to our homes, individual house supply cables are connected to one supply phase and neutral, so you may be without power while your neighbour, connected to a different supply phase, may still have it. Report any power supply interruption to your local distribution company's 24-hour emergency number (listed under 'Electricity' in your telephone directory). They will confirm whether there is a power cut in your area, and give you an idea of when power is expected to be restored.
3.  If there is no power cut and you are the only property without power, there is either a fault on your home's main supply cable or your main service fuse has blown. Call your distribution company's emergency number, and ask for an engineer to call round and rectify the fault.

Note  RCDs occasionally suffer from what is called nuisance tripping, caused by a transient fault on the supply system, but they can usually be reset immediately. If you suffer from constant nuisance tripping, report it to your electricity distribution company.

---

**What you need:**
- fuse wire *or* cartridge circuit fuse
- electrical screwdriver
- side-cutters
- wire strippers
- replacements for damaged accessories
- replacement circuit cable
- junction boxes

**CHECK 2**
- the cause of sparks or a burning smell emanating from any fixed wiring accessory. Turn off the power, remove the face-plate and examine it for loose connections or signs of overheating. Remake connections, and replace accessories if damaged

**TIP 2**
If you or a member of your family receives a shock from an appliance, isolate it immediately for testing and repair. If a shock is received from any part of the fixed wiring, track the fault down immediately or call a qualified electrician to locate and repair it for you. A repeat could kill.

# Mending Fuses

Fuses are safety devices fitted at various points on your wiring system to protect appliances and cables from damage in the event of a fault. Every circuit has one (although on modern installations miniature circuit breakers – MCBs – may take their place), and modern rectangular-pin plugs contain one as well. Each is an intentionally weak link, designed to fail if too much current is drawn by the circuit or appliance the fuse is protecting.

There are two types of circuit fuse. Rewirable fuses have a length of fuse wire of the appropriate rating fitted between two terminals on the fuse-holder. They are cheap to replace and it is easy to see when a fuse has blown, but they are easy to abuse; worst of all, the overload protection they offer is poor because it takes twice the current of the fuse wire rating to blow it.

Cartridge fuses contain an enclosed cartridge similar to (but larger than) the ones used in fused plugs. They cost more than rewirable fuses but are easier to replace.

Since different fuse ratings are different sizes, you cannot fit the wrong one in an individual fuse-holder. Furthermore, they offer better overload protection because the current rating needed to blow the fuse is only 1½ times the fuse rating. By comparison, an MCB trips when the overload current exceeds just 1¼ times its stated rating.

## What to do

**Option 1** With rewirable fuses, switch off the power, remove the fuse-holder and loosen the terminals so you can remove the old fuse wire. Then fit new wire between the terminals (Figs 211 and 212), replace the fuse-holder and restore the power.

**Option 2** With cartridge fuses, turn off the power and remove the fuse-holder as before. Then take out the old cartridge and fit a replacement of the correct rating (Figs 213 and 214).

What you need:
- fuse wire *or* cartridge fuse of the appropriate rating for the circuit concerned
- electrical screwdriver
- side-cutters
- continuity tester for checking cartridge fuses

CHECK
- that you *always* have a stock of spare fuse wire or cartridge fuses of the appropriate ratings

TIP
If you do not have a continuity tester, use a torch with a metal case to check whether a cartridge fuse has blown (*see* Check 1, page 73).

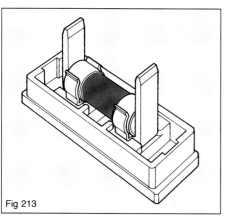

Fig 211

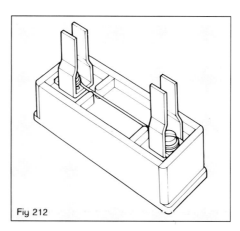

Fig 212

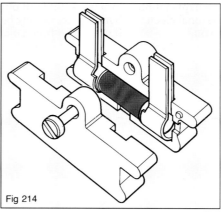

Fig 213

Fig 214

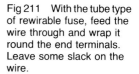

Fig 211 With the tube type of rewirable fuse, feed the wire through and wrap it round the end terminals. Leave some slack on the wire.

Fig 212 With the bridge type, lay the wire slackly across the ceramic bridge.

Fig 213 Some cartridge fuses fit into a spring clip between the pins.

Fig 214 In others you have to separate the two halves of the fuse-holder to release the fuse.

# Replacing Pendant Flex

There are several reasons why you might want (or need) to replace the flex that links a pendant lamp-holder to its ceiling rose. It may have become discoloured with age (white flex can go yellow, especially in smoky atmospheres). You may want a longer flex, for example to lower a light over a dining table. It may even have suffered damage that has caused one of the cores to part company – this is common when too heavy a lampshade has been fitted on the lamp-holder (see Check for details of the maximum weight flexes can support).

It is important to fit the correct type of flex. If the lamp-holder is plastic, use two-core round flex. However, if the lamp-holder is brass, you must use round three-core flex with an earth core to earth the lamp-holder to the earth terminal in the ceiling rose.

If you are planning to fit an enclosed lampshade or use a high-wattage bulb in the lamp-holder, it is a good idea to fit a heat-resistant lamp-holder.

## What to do

Start by turning off the power to the circuit supplying the light (see page 34). It is not sufficient to turn off the light switch. Unscrew the rose cover and undo the terminal screws that secure the flex cores to the rose. Lift the lamp-holder and flex down, unscrew the lamp-holder cover and disconnect the flex cores.

Cut a new length of flex of the appropriate type, and strip about 50mm (2in) of the sheath from one end. Prepare the cores and connect them to the lamp-holder terminals. Loop the cores round their anchors and fit the lamp-holder cover. Then strip 100mm (4in) of sheath from the other end of the flex, and again prepare the cores.

Thread the flex through the rose cover, then connect the cores to the rose terminals and loop them over the flex anchors. Finally replace the rose cover.

Where the flex connection is made within a conduit box (Fig 217), link the new flex to the mains via the strip connector.

**What you need:**
- replacement two-core or three-core flex
- electrical screwdriver
- side-cutters
- handyman's knife
- wire strippers
- new heat-resistant lamp-holder

**CHECK**
- that the flex can support a heavy lamp shade. The maximum weight each size of flex can safely carry is:

$0.5mm^2$  2kg (4½lb)
$0.75mm^2$ 3kg (6½lb)
$1mm^2$    5kg (11lb)

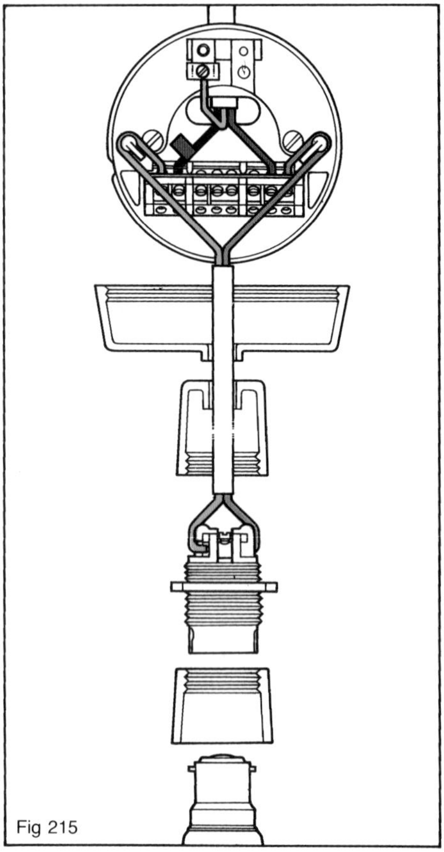

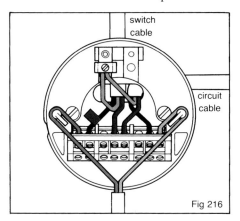

Fig 216

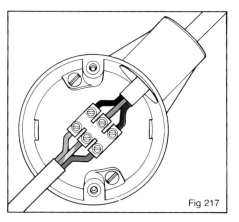

Fig 217

Fig 215 (bottom left illustration)

Fig 215  How the components of a pendant lamp-holder and ceiling rose fit together. Connect the new flex to the lamp-holder terminals first, loop the cores over the flex anchors and fit the lamp-holder cover. Then thread the flex through the rose cover and connect the flex cores to the rose terminals. Again loop the cores over the flex anchors before replacing the rose cover.

Fig 216  With loop-in wiring, the flex cores still go to the end terminals.

Fig 217  Where the light is wired via a conduit box, the new flex is linked to the mains via a strip connector. The other end of the flex goes to a terminal block within the light fitting.

# Replacing Flex on Appliances

The flex on many portable appliances, especially those in constant use such as electric kettles, irons, food mixers and hair driers, gets a lot of wear and tear and is prone to accidental damage too. So too is the flex on appliances such as heaters, where the flex trails across floor surfaces and may be damaged by feet or furniture. If the flex sheath is broken, it is easy for the flex cores inside to be exposed, and there is obviously a potential danger that someone using the appliance could touch a live wire.

A damaged flex sheath should be repaired immediately with a bandage of PVC insulating tape to make it temporarily safe, and should be replaced with new flex at the earliest opportunity. Always check that you are using flex of the right type and current rating (*see* Check).

It is well worth carrying out a regular inspection of the condition of the flex on all your portable appliances, so that any damage can be spotted and repaired before someone receives a shock.

## What to do

Unplug the appliance from the mains, and locate the screws securing its outer casing. As you remove them, tape them to the casing next to their holes so you put the right one back in each hole – they are often of different lengths, and mixing them up can cause problems when you come to replace them.

On most appliances it is easy to find the terminal block to which the flex cores are connected. Before you disconnect any of them make a note or draw a simple sketch to show how the existing cores are connected. Note too whether the flex passes through a rubber grommet as it enters the appliance casing, and whether it is secured in a cord grip.

You can now disconnect the old flex cores, release the cord anchorage if fitted and draw the flex out of the appliance. Cut a length of new flex of the appropriate type and rating (*see* Check), and prepare the cores using the old flex as a pattern; the earth core, for example, may have to be longer than the others to reach its terminal on the appliance casing.

Feed the flex through the grommet if one is present, and lead it to the appliance terminal block. Refer to your notes, and connect each core to its appropriate terminal. Make sure that all cores are securely connected. Then trap the flex sheath in the cord grip, leaving a little slack within the appliance, and replace the casing using the screws you removed earlier. Complete the job by wiring the plug to the other end of the flex, and check that the plug fuse is the correct size for the appliance (*see* page 22).

**CHECK**
- that you use flex of the right type and size for the appliance being rewired. Use flat or round two-core flex for double-insulated appliances marked with the double square ▢ symbol, and round three-core flex for other appliances. Choose the flex size to match the appliance wattage:

0.5mm$^2$ up to 720W
0.75mm$^2$ up to 1.4kW
1mm$^2$ up to 2.4kW
1.5mm$^2$ up to 3.6kW

**TIP**
On braided flex, wrap PVC tape over the cut braid to stop it from fraying.
  Take this opportunity to fit a longer flex to the appliance if it needs it.

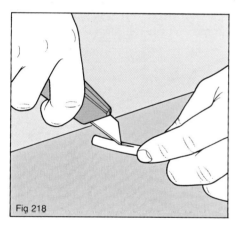

Fig 218

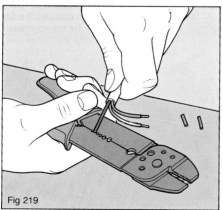

Fig 219

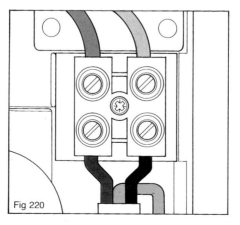

Fig 220

Fig 218  Slit the new flex sheath carefully with a knife and cut off the excess.

Fig 219  Use wire strippers to remove about 12mm (½in) of the core insulation.

Fig 220  Make a sketch of which core goes where before you disconnect the old flex.

# Extending Flex Safely

Many electrical appliances are sold with surprisingly short flexes attached, and even in a house which has plenty of conveniently-sited socket outlets it is often difficult to be able to use the appliance where you want to. The ideal solution is to replace the existing flex with a longer one (see page 77), but in some cases this is not possible or desirable. For example, you may need a long flex to enable your lawn-mower to reach the bottom of the garden, but having it permanently attached to the machine could make it awkward to store neatly.

The solution is to extend the flex by attaching an extra length. Never twist the cores of the old and new flex together and cover the join with tape; instead always fit a special flex connector. Use a one-piece connector for a permanent extension, and a two-part one for situations where you will want to disconnect the extension flex for ease of storage. For indoor use, fit flex of the same type as that on the appliance. For outdoor use, fit white or orange flex.

## What to do

Start by deciding what length of extra flex and which type of connector you need. Buy flex of the appropriate type and size for the appliance (see Check on page 77); use two-core flex with double-insulated appliances, three-core flex otherwise. Then buy the appropriate type of connector containing either two or three terminals. Remove the appliance plug and prepare the new flex cores.

**Option 1** If you are using a one-piece connector, unscrew the cover, pass the end of each flex through the cord grip and connect the cores to the terminal tubes, like to like. Anchor the cord grips and replace the cover.

**Option 2** If you are using a two-part connector, start by attaching the flex from the appliance to the plug part of the connector. Open the connector, feed the flex through the cover and cord grip and connect it to the terminals. Repeat the operation to connect the mains lead to the socket part of the connector.

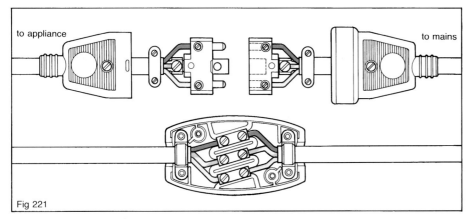

Fig 221

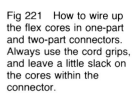

Fig 221   How to wire up the flex cores in one-part and two-part connectors. Always use the cord grips, and leave a little slack on the cores within the connector.

## Extensions for Outdoor Use

If you plan to use an extended flex out of doors, make sure that the connectors are splashproof and made from rubber – not rigid plastic which could crack if damaged or trodden on.

If you have a range of garden power tools – lawn-mowers, hedge-trimmers, lawn rakes and the like – you can fit a connector plug to each one, and then plug it into the socket of your extension lead.

Always plug appliances used out of doors into a socket with RCD protection (see page 57 for more details).

## Using Extension Reels

Most do-it-yourselfers own an extension reel – a length of flex coiled up on an enclosed reel with one or two socket outlets mounted on the reel body. You do, however, need to be careful when using them.

Firstly, they should *always* be fully uncoiled before use. Otherwise, the coiled flex may overheat, causing the insulation to melt and resulting in a short circuit. Secondly, you should check the flex rating before using the reel with high-powered appliances such as heaters. Many extension flexes are only rated at 5 amps.

# Repairing Damaged Cables

However careful you are, there may come a time when you drill through a hidden cable in a wall or underneath the floorboards. It is to help prevent this that cables in walls should always be run vertically or horizontally; then at least you know where cables are likely to run, and can avoid drilling a hole that might hit one. For the same reason, cables crossing floor voids should run through holes drilled in the joist centres, rather than in notches cut in their top edges.

Unfortunately, not everyone follows these rules, and if previous occupants have done any wiring work on your home the unexpected can always happen. If you do hit a cable, the drill bit may pierce just one core, or may cause a short circuit by drilling through several. There will probably be a bang and a flash, and the circuit fuse or MCB will cut off the power to the circuit. You then have to repair the damage caused to the circuit cable before you can restore the power to the affected circuit.

## What to do

You will probably have had quite a fright when you hit the cable, although the double insulation on your drill will have prevented you from receiving an electric shock. When you have recovered your composure you can start repair work.

**Option 1**   Cables in walls. If you have hit a cable run in an unprotected chase in the plaster, you must trace the run up or down the wall. You then must cut away the plaster and remove the cable run from the accessory it served to the ceiling or floor void. Run new cable from the accessory to the void, and cut back the damaged cable at a convenient point close to a joist. Prepare the cable ends and reconnect them within a three-terminal junction box (Fig 224), connecting like cores to like. Fit the junction box cover and restore the power. Then secure the new cable in the chase with cable clips at about 300mm (1ft) intervals, and plaster over the chase containing the cable to complete the repair.

If the cable ran in a conduit and there is sufficient room within the conduit to accommodate a strip connector, it may be possible to cut away part of the conduit so you can gain access to the cable. Then cut through the cable, strip the cores and reconnect them at the strip connector. Push this back into the conduit, restore the power and plaster over the hole.

**Option 2**   Cables under floors. If you hit a cable run in a notch cut in the top of a joist, lift the floorboards, cut through the cable at the point of damage, and prepare the cable cores as before. Reconnect them at a three-terminal junction box screwed to the side of the joist.

**What you need:**
- three-terminal junction box
- cold chisel
- club hammer
- floorboard chisel
- side-cutters
- handyman's knife
- wire strippers
- electrical screwdriver
- earth sleeving

**CHECK**
- that bare earth cores are sleeved before being connected within junction boxes

**TIP**
If you find a cable running in an unexpected direction, mark where it goes on floorboards or on undecorated walls as a reminder.

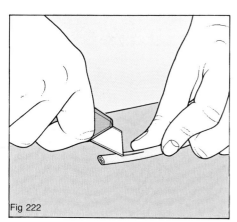

Fig 222

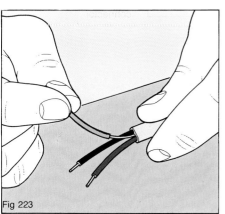

Fig 223

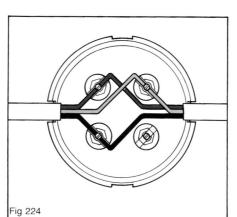

Fig 224

Fig 222   Cut the cable at the point of damage, slit the sheath and cut off the excess.

Fig 223   Prepare the cable cores for reconnection, adding PVC sleeving on the bare earth core.

Fig 224   Reconnect the cable cores within a three-terminal junction box.

# Replacing Damaged Accessories

Wiring accessories with plastic face-plates are easily damaged, especially if they are surface-mounted and are positioned near floor level where they can be knocked by furniture being moved or by boisterous children riding pedal cars round the house. Such damage can result in live cores or terminals being left exposed if part of the face-plate or mounting box breaks away, and this is obviously highly dangerous if someone touches them.

If such damage does occur, carry out a temporary repair immediately (see below) to make the accessory safe, and avoid using it until such time as a replacement can be fitted. Buy and fit the new accessory as soon as possible.

You may also need to replace an accessory such as a ceiling rose because it has been sealed up with carelessly applied paint over the years and you are unable to unscrew the rose cover as a result. This will require some relatively unsubtle demolition work.

## What to do

Start by turning off the power to the circuit supplying the accessory (see page 34); never try to replace an accessory with the power on. Undo the two screws securing the face-plate to its mounting box and ease the face-plate away carefully. Loosen the screws securing the cores to their terminals so you can disconnect the face-plate, and discard it. Then reconnect the cores to the new accessory, fold the cable(s) back neatly into the mounting box and attach the face-plate.

If you have a ceiling rose cover that is sealed to its base-plate with paint, you have no option but to break it to get it off. Turn the power supply to the rose off, and use a small cold chisel and a hammer to crack the edge of the rose cover so you can prise it off. Then disconnect the flex and cable cores from the base-plate, unscrew it, discard it and replace it with a new rose (see page 37).

**What you need:**
- replacement accessory
- replacement surface-mounting box
- screwdriver
- electrical screwdriver
- handyman's knife
- cold chisel
- hammer
- PVC insulating tape

**CHECK**
- that the power to the circuit is off before you try to fit a replacement accessory

**TIP**
When removing a flush face-plate or a surface-mounted box, cut round it with a knife first to free any paint bond and to avoid lifting and tearing wall-coverings.

When fitting a replacement accessory, use the old face-plate fixing screws. The new ones may have a different thread pattern to that in the fixing box lugs, making them impossible to locate.

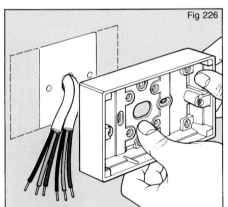

## Temporary Repairs

If an accessory face-plate, a plug or even the plastic casing of an appliance is cracked, you can make a temporary repair with PVC insulating tape. On an accessory, lay two or three strips over the damaged area to prevent anyone touching live parts, and if possible do not use it until it can be replaced. On plugs, wrap tape round the plug body, passing it between the plug pins (Fig 227). On an appliance, pass tape right round its body. Do not allow temporary repairs to become permanent; get new parts fitted as soon as possible.

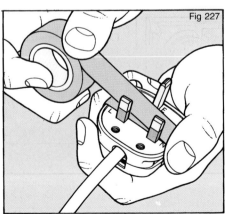

Fig 225  Replace damaged accessories at the earliest opportunity.

Fig 226  If a surface-mounted box is damaged, replace that too.

Fig 227  Use PVC insulating tape to make temporary repairs to cracked plugs and other fittings.

# Repairing Cord-Pull Switches

Ceiling-mounted cord-pull switches are widely used in bathrooms for safety reasons; an ordinary plate-switch must be positioned at least 2m (6ft 6in) from a bath or shower, and many modern bathrooms are so small that this requirement cannot be met. They are also useful elsewhere in the house, particularly for controlling bed-side lights. The pull cord is actually in two sections; a short length emerging from the body of the switch, linked to the main cord and its bell-shaped pull by a small threaded two-part connector.

The cord pull is the weak spot. Since it gets constant use, it can eventually wear, fray and finally break. If the break is below the connector, it is a simple matter to unscrew this and fit a replacement length of cord; this is available, complete with a new cord pull, from all electrical shops. However, if the break is between the connector and the switch itself, the switch must be disconnected from the mains so the repair can be carried out.

## What to do

If the break is in the lower section of the cord, simply unscrew the connector and pull the section of broken cord out. Cut the new cord to the required length, thread it through the connector and the cord pull and knot the ends. Reattach the lower part of the connector to the upper one.

If the break is in the upper section, start by turning off the power supply to the switch (see page 34). Then unscrew it from its backing plate, and disconnect the cable cores so you can repair the switch on your work-bench.

Undo the screws holding the terminal plate to the switch, while holding it with your other hand to stop the spring inside from firing bits everywhere. Lift it off carefully to expose the top of the cord anchor, and pull out the old cord. Feed a new length in and knot its end, then re-place the terminal plate and reconnect the switch to the mains supply.

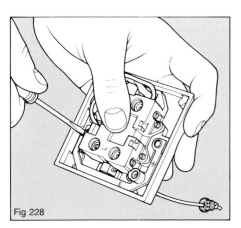

Fig 228

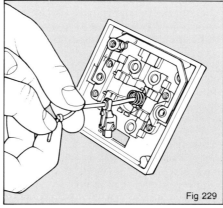

Fig 229

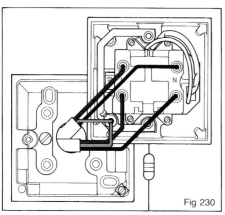

Fig 230

## Other repairs

Ceiling-mounted switches can suffer from failed fixings, caused by people pulling the cord too sharply. If the switch starts to pull away from the ceiling, repair the fixings before the whole switch comes loose. Turn off the power and unscrew the switch body. Then remove the fixing screws from its base-plate so you can see what has failed. Using longer fixing screws may make a firmer fixing; otherwise you may have to reposition the switch slightly so you can make new fixings into the solid timber of a joist or into a batten fixed between adjacent joists.

Fig 228 Disconnect the switch from its backing plate and hold the terminal plate in place while you undo the fixing screws. Lift it off carefully so the spring inside does not fly out.

Fig 229 Draw out the old cord and feed a new piece through the cord anchor. Knot the end and cut off any excess.

Fig 230 Reconnect the circuit cables to the switch, making sure the supply cable goes to the 'feed' terminals and the appliance cable to the 'load' terminals.

# Basic Appliance Repairs

Many domestic appliances, especially those the trade calls 'white goods' – things like fridges, freezers and washing machines – are too complicated for the average do-it-yourselfer to repair when something goes wrong. The fault is usually the failure of a major component such as a motor or heater, or else a malfunction in the increasingly complex control wiring, and is best left to a qualified repair agent. However, many smaller appliances have components that wear out with use – not necessarily as the result of a fault – and it is often possible to replace these yourself. However, if the appliance is still under guarantee you may invalidate the terms of the guarantee by attempting to do-it-yourself repair. Check this point before you tackle anything.

There are three broad appliance groups: those with a heating element, those with a motor and those with both a heating element and a motor.

## General Repairs

If an appliance stops working, check its plug fuse, the connections within the appliance and the plug, and the flex continuity (see page 73).

Many appliances simply get clogged up with dust, fluff, hair and grease, and cleaning them can dramatically improve their performance. Examples include vacuum cleaners, fan heaters, hair driers and extractor fans. Disconnect the appliance from the mains, then undo the screw securing the casing so you can remove it. You can then pick out large pieces of waste matter by hand. Use a dry paintbrush to remove dust and fluff from heating elements, and a soapy cloth to remove grease from extractor fan blades. Replace the casing.

## Electric Heaters

Fires with radiant elements eventually suffer from element failure. With many fires it is simple to remove the failed element and fit a replacement. Unplug the fire and remove the guard or cover plate to gain access to the element. Then unscrew its fixing nuts or prise it out of its holder and take it with you when you go shopping for a replacement. It helps to take a note of the fire make and model number with you too.

Repairs to the heating elements of fan convectors should be left to an expert.

## Electric Irons

As with fires, it is the heating element which burns out in electric irons. You can usually replace this on an older non-automatic iron, by removing the cover and undoing the terminal screws to release the element contact strips. Then remove the nuts holding the pressure plate on, and lift it off to reveal a heat-resisting pad and the element itself. Replace it and reverse the dismantling sequence to reassemble things.

Have automatic and steam irons expertly repaired.

## Electric Kettles

Once again, element failure is the commonest fault with electric kettles. Repair is relatively straightforward on many non-automatic kettles; you simply unscrew the collar holding the element to the kettle body and lift the old element out. Buy new sealing washers when you buy the replacement element, and fit the rubber one inside the kettle and the fibre one outside as you fit the new element to ensure a water-tight seal.

Automatic kettles with thermostatic control need expert repair.

## Appliances with Motors

Many appliances have motors in which power is picked up from the spinning commutator by small carbon brushes. Examples include vacuum cleaners, food mixers, hair driers and power tools. These brushes wear out eventually, leading to sparking and intermittent operation of the motor. If you can get access to the motor, replacing the brushes themselves is usually relatively easy to do.

Buy appropriate replacement brushes for the appliance concerned from a service agent. Then locate and remove the brush caps, lift out the old brushes and blow any dust out of the brush tube. Place a new brush in each tube, depress the retaining springs and replace the brush caps. Replace the appliance casing if it was removed for ease of access, and test the operation of the motor.

---

**What you need:**

- replacement elements
- assorted screwdrivers
- cleaning tools
- replacement brushes

**CHECK**

- that the appliance is unplugged or disconnected from the mains before you attempt to carry out any repairs
- that you will not invalidate any guarantee by attempting a repair

# EXPENSIVE JOBS

All the projects described in this book are well within the capacity of any competent, careful and intelligent do-it-yourselfer who has a reasonably recent wiring system to work on. However, there are still many old and sub-standard house wiring systems in existence, and as mentioned earlier any attempts to extend an already potentially overloaded system that may also be in a very poor state of health is asking for trouble.

If you have an old wiring system such as this, the first thing you should do is to call in a professional electrician to check that it is safe to use. Only if it is billed as safe should you contemplate making any extensions or alterations to it, and even then you should proceed with caution. What you really need is a complete re-wire – new circuit cables, new accessories everywhere, plus a new consumer unit.

## Tell-Tale Signs

If your wiring is old, you are probably only too well aware of the fact. For a start, you will have realized that you do not have nearly enough socket outlets around the house to cope with the demand of today's multitude of electrical appliances. You may still have socket outlets with round holes, designed to take old-fashioned round-pin plugs – a sure sign that the system is at least forty years old and seriously out of date. Even if you have modern socket outlets, don't take that as proof that your house has modern wiring. It is a popular dodge – and a particularly dangerous one – for sellers to fit new sockets and switches to impress (and deceive) would-be purchasers that the house has been rewired when it has not.

If you have ever noticed a 'hot' smell when you are using an appliance, it is probably caused by overheating – either within a socket or switch, due to poor contacts causing sparking, or else somewhere hidden deep within the wiring. No wiring accessory should ever feel warm to the touch.

You will probably also suffer from a lot of unexpected blown fuses – another indication that all is not well and that short circuits or current leaks are occurring at various points round the system due to bad connections or failed insulation.

What goes on at the fuse-board gives an excellent indication of the health of your wiring. If you find an assortment of individual metal fuse-boxes, perhaps containing old porcelain fuse-holders and interlinked by a spaghetti-like sprawl of cables, it is high time a new consumer unit was fitted to take its place. This is especially true if you find that individual circuits have two fuses each; such double-pole fusing is no longer used for house wiring and could be in a highly dangerous condition. Immediate replacement is absolutely essential.

Look too at the individual circuit cables at the fuse-board; if they have rubber insulation rather than grey or white PVC, you may well find that the insulation is brittle. This indicates that your insulation may have failed just about anywhere on the system – this is like a time bomb waiting to start an electrical fire.

All this adds up to a virtually terminal diagnosis. Any system in (or approaching) this condition must be rewired at the earliest opportunity; it is too late for any remedial work. It is also literally a matter of life and death – an old plumbing system will just make a mess, but an old wiring system can kill.

Fig 231 (*above*)  Jobs such as fitting a new consumer unit in place of an array of old fuse-boxes should be left to a professional electrician.

> **SYSTEM CHECK-LIST**
> If your system has any of the following features, you probably need a re-wire:
>
> - round-pin plugs and socket outlets
> - round metal or bakelite light switches set on wooden blocks
> - wiring run in metal conduit
> - separate main switch and fuse-boxes
> - cables with rubber sheathing

Rewiring a house is a job for professional electricians. Their knowledge is essential in ensuring that the system is correctly designed and installed to meet the requirements of the wiring regulations. They will also do it much more quickly than an amateur could, an important point when you contemplate life without an electricity supply. However, there is no reason why you should not be closely involved in planning the sort of system you want.

## Making Plans

If you are going to the trouble of having a complete re-wire, now is a golden opportunity to plan for the future and create a wiring system that will not only cater for your likely future needs, but which will also be much easier to alter and extend than a traditional system. Here are some points to consider.

The consumer unit Its size will be dictated by the number of individual circuits you want. The basic needs of a modern home are three lighting circuits, three ring main power circuits, plus separate circuits to an immersion heater and an electric cooker; fit the latter even if you cook by gas, in case a future occupant prefers to cook electric. It also makes sense to provide for a sub-circuit to supply power to an outbuilding, even if you do not need it at the moment. Once you have worked out your present requirements, choose a unit with at least two additional spare fuseways to allow for extra circuits to be added in the future with the minimum of disruption.

Socket outlets Always specify double-socket outlets, unless there is an overriding reason why a single outlet should be installed at a particular location. In places like kitchens and living rooms where you will use a lot of individual appliances, it may even be worth specifying triple outlets. Remember that there is no limit to the number of outlets each circuit can serve, so you may as well have too many rather than too few.

Think also about the siting of your outlets. They do not all have to be at skirting board level, and there are many situations where waist-level sockets are much more practical.

Lighting points The gradual introduction of plug-in luminaire support couplers (see page 39) means that you can have far greater flexibility in planning your lighting. With LSCs you can provide lighting sockets on walls and ceilings in every room, even if they are not all needed at the present; they can simply be left blanked off until they are. Plan to use fittings with modern compact fluorescent lamps rather than conventional tungsten ones; you will get far longer bulb life and also lower electricity bills.

Think also about the possibilities of low-voltage lighting, which allows you far greater flexibility (and far more lighting points) than mains-voltage fittings.

How the wiring is run Perhaps the most important departure from tradition that you can consider is how you actually wire the house up. It is actually rather silly in these days of fitted carpets and built-in furniture to run cables under floorboards or buried in walls where they are impossible to get at without serious disruption to the room's decorations.

Modern trunking systems mean that all the wiring traditionally run beneath the floor can be placed instead in neat, unobtrusive skirting run round the room. This takes care of wiring for socket outlets at skirting board level, and means that any alterations to the system can be carried out with the minimum of disruption. You will also have skirtings that never need painting again!

Architrave trunking systems similarly allow you to run cables to light switches by room doors. That just leaves the problem of getting cables to accessories mounted on ceilings or away from wall perimeters. Here, it makes sense to run all the cables in conduit, and to place a nylon drawstring in each conduit run so that future alterations can easily be carried out by pulling new cables through the existing conduit. How's that for forward planning?

CHECK
- that you employ only qualified electricians – ideally members of either the Electrical Contractors' Association (ECA) or the National Inspection Council for Electrical Installation Contracting (NICEIC). You can get lists of ECA and NICEIC members working in your area from your local electricity distribution company's showroom
- that your contractor carries out tests on the new system for verification of polarity, effectiveness of earthing, insulation resistance and ring circuit continuity. He should issue you with a test certificate when the work is completed

Note If you carry out major alterations to your system such as adding new circuits, you should contact your local electricity distribution company and ask for an application form for approval of the new wiring. The company will then arrange to connect the wiring to the mains and will almost certainly carry out tests to ensure that the work complies with wiring regulation requirements.

# FACTS AND FIGURES

This section is intended as a handy reference guide to the range of electrical accessories and materials you are likely to need for various wiring jobs around the house and garden. It will help you to see at a glance what is available, and what to use where, so you can plan your requirements in detail and draw up shopping lists for individual projects – essential for keeping job costs under control.

It also contains a summary of the most widely used wiring layouts, for both lighting and power circuits, so you can refer quickly to them while you work on your wiring instead of having to leaf backwards and forwards between subjects elsewhere in the book.

Lastly, on page 94 there is a detailed glossary of all the terms used in the book, plus a list of useful addresses of product manufacturers and trade associations.

## Cable and Flex

Below is a check-list to help you choose cable and flex of the right type and specification for each job.

**Two-core and earth cable**  Use PVC-sheathed and insulated two-core and earth cable for all general circuit wiring. The sizes you need are:

$1mm^2$ – for lighting circuits protected by a 5-amp circuit fuse or MCB.
$1.5mm^2$ – for power circuits to appliances rated at under 3.6kW protected by a 15-amp fuse or MCB.
$2.5mm^2$ – for ring circuits protected by a 30-amp fuse or MCB, for radial circuits protected by a 20-amp fuse or MCB, and for all spurs taken off such circuits.
$4mm^2$ – for radial circuits protected by a 30-amp fuse or MCB, for spurs taken off such a circuit, and for power circuits to appliances rated at up to 7kW.
$6mm^2$ – for power circuits to appliances rated at up to 8.4kW protected by a 45-amp fuse or MCB.
$10mm^2$ – for power circuits to appliances rated at up to 11.5kW protected by a 45-amp fuse or MCB.

**Three-core and earth cable**  Only one size of this cable is commonly used:

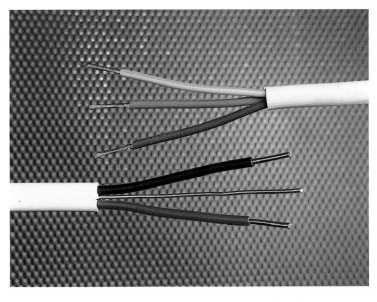

$1.0mm^2$ – for two-way switching arrangements on lighting circuits.

**Single-core cable**  The following sizes are available:

4, 6, 10 or $16mm^2$ – with green and yellow insulation for earthing and cross-bonding connections.
16 or $25mm^2$ – with red or black insulation for use as meter tails and for connecting consumer units and separate switch-fuse units via a distribution box.

**Circular three-core PVC-sheathed and insulated flex**  This is used for all appliances requiring an earth, and for metallic lamp-holders.
**Circular two-core PVC-sheathed and insulated flex**  This is used for all double-insulated appliances and non-metallic lamp-holders.
**Flat two-core PVC-sheathed and insulated flex**  This is used as an alternative to round flex for portable double-insulated appliances.
**Two- and three-core braided flex**  This is used for appliances such as electric heaters and is used exclusively indoors.
**Unkinkable three-core flex**  This is used for hand-held appliances such as irons and kettles.

Fig 232 (*above*)  Cable and flex are your basic raw materials for all wiring work. Red cable cores and brown flex cores are live, black cable cores and blue flex cores are neutral.

# Power Circuits

You need a wide variety of wiring accessories for the various power circuits in the house, ranging from the commonplace socket outlets to more exotic accessories such as shaver units and cooker controls. Below is a summary of the various types of accessories available, with details of the varieties they come in and what depth of mounting box each will require.

You are likely to buy just one or two examples of most of the accessories listed here. However, remember that you can make substantial savings on the more widely used accessories such as socket outlets by buying in bulk packs of ten.

Fig 233   Sockets and switches.

## Socket outlets

**Standard single socket outlet**  This is about 85mm square, and fits over a single mounting box 25mm or 35mm deep (the deeper box allows more room for the circuit wiring). It is available in three versions – unswitched, switched and switched with a neon indicator light.

**Double socket outlets**  These measure approximately 145 x 85mm, and again fit over boxes 25 or 35mm deep and come in the three versions described for single socket outlets.

**Triple socket outlets**  These measure approximately 200 x 85mm, and again need boxes 25 or 35mm deep. Usually only switched versions are available.

**RCD socket outlets**  These are the same size as a double socket outlet. Most require a box 35mm deep, although some brands will fit shallower 25mm boxes.

## Fused Connection Units

The standard fused connection unit (FCU) is the same size as a single socket outlet – about 85mm square – and generally needs a mounting box 25mm deep. There are three main types. The first allows flex entry through the front of the accessory face-plate, the second through a notch in the face-plate's edge. The third is designed for use in wiring fused spurs, and allows for incoming and outgoing cables.

FCUs of each type are available in switched or unswitched versions, with or without a neon indicator light. The recessed fuse carrier is fitted with a cartridge fuse of the type used in standard 13-amp fused plugs; the amperage of the fuse should be selected according to the wattage of the appliance it is serving, or the spur it is supplying.

Two separate FCUs can be mounted side by side on the same box if required, using a special dual mounting box.

## Double-Pole Switches

**20-amp double-pole (DP) switches**
These are the same size as a single socket outlet or FCU, and need a 25 or 35mm deep mounting box. There are two types, one allowing flex entry through a hole in the face-plate and the other designed to accept an outgoing cable. Both are available with or without neon indicator lights. Some manufacturers offer switches with the face-plate labelled 'water heater', and also a special dual switch used for controlling twin element immersion heaters.

**Higher-rated DP switches**  These are available with ratings of 32, 45 and 50 amps. The 32-amp version is the same size as a single socket outlet, and needs a 35mm deep box. Most 45- and 50-amp versions are the same size as a double socket outlet – about 145 x 85mm – and also need a 35mm deep box, but single-socket size 45-amp switches are also available. Some types come with a neon indicator light, and all are usually labelled to show the appliance they control.

# Power Circuits

## Ceiling-Mounted Switches

Ceiling-mounted DP switches come in two ratings – 15-amp or 16-amp and 40-amp or 45-amp, both with or without a neon indicator light. Most are intended for surface-mounting, but can also be mounted over standard recessed conduit boxes (round switches) or on 35mm deep flush metal boxes (square types). The higher-rated type must have an on/off 'flag' indicator so the user can see whether the circuit from the switch is on even if the neon indicator light fails.

## Cooker Controls

**Cooker control units** Those with an integral 13-amp socket outlet are an alternative to a plain 32-amp or 45-amp DP switch for controlling free-standing or built-in cookers, ovens and hobs. Some manufacturers offer a version the same size as a double socket outlet which can be fitted on a 35mm deep box, but most cooker control units are a little larger than this – commonly around 160 x 115mm – and may require a deeper 47mm box to provide room for the heavy-duty circuit wiring. The unit has separate switches for the socket and cooker, and is available with or without neon indicator lights.

**Cooker connection units** These enable the cooker to be connected to the mains wiring. They are the same size as a single socket outlet, and are generally fitted to 25mm deep mounting boxes.

## Shaver Units

**Shaver socket outlets** These are for use in rooms other than bathrooms and wash-rooms, are the same size as a single socket outlet, and fit on a 25mm deep flush or surface mounting box.

**Shaver supply units** These contain a transformer and are intended for use in bathrooms. They are the same size as double socket outlets (mounted vertically), and need a 47mm deep box.

**Shaver striplights** These are surface mounted over a recessed connection box.

## Aerial Sockets and Telephone Outlets

**TV/FM radio aerial sockets** These are the same size as single socket outlets, and fit on 25mm deep mounting boxes. Versions are available with one or two outlets, and the latter can supply two TV/FM connections or separate UHF/VHF television and VHF/FM radio connections.

For multiple aerial applications, two sockets can be mounted side by side over a double mounting box.

**Telephone outlets** These are also single socket size, and again fit on 25mm deep mounting boxes. Smaller surface-mounted outlets about 70mm square come complete with a plastic mounting box.

## Junction Boxes

**Junction boxes** Those for power circuits are either round, for mounting beneath floorboards, or rectangular for surface mounting. Both types are rated for power circuit use, and contain three separate terminals which are usually capable of accepting up to four 4mm$^2$ cable cores.

**Strip terminal blocks** These are used for providing terminal connections within power-circuit mounting boxes, and are designed to be mounted between the fixing lugs of standard flush metal or surface-mounted plastic boxes. A blank cover plate is then fitted to conceal the terminals.

## Conduit and Mini-Trunking

**Round PVC conduit** This is available in a range of diameters, but the 20mm size is the most widely used on domestic installations in conjunction with conduit boxes (see page 91 for more details).

**Oval-section conduit** This is for cable runs to light switches and other wiring accessories and is also available in several sizes. The 13 x 8mm, 16 x 10mm and 23 x 11mm sizes are the most widely used for domestic use.

**Mini-trunking** This is used for surface wiring and comes in sizes from 16mm square up to 50 x 32mm. Most types are white, and some have a self-adhesive backing for easy fitting.

## Ring Circuit Check-List

1.  The circuit is wired from a 30-amp fuseway in the consumer unit, to which both live conductors are connected at the ends of the circuit cable.
2.  The circuit is wired throughout using 2.5mm² two-core and earth PVC-sheathed and insulated cable.
3.  The circuit can supply an unlimited number of individual socket outlets and fused connection units (FCUs) on the main ring circuit.
4.  The circuit can supply as many spurs as there are socket outlets or FCUs on the main ring circuit.
5.  Spurs must supply only one single, or one double socket outlet, or one fused connection unit (formerly, each spur could supply two outlets).
6.  A socket outlet or junction box on the ring circuit can supply only one spur.
7.  One ring circuit supplies sockets and FCUs in rooms with a maximum floor area of 100 sq m (1,080 sq ft).

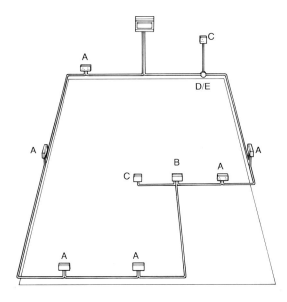

Fig 234   Ring circuit.

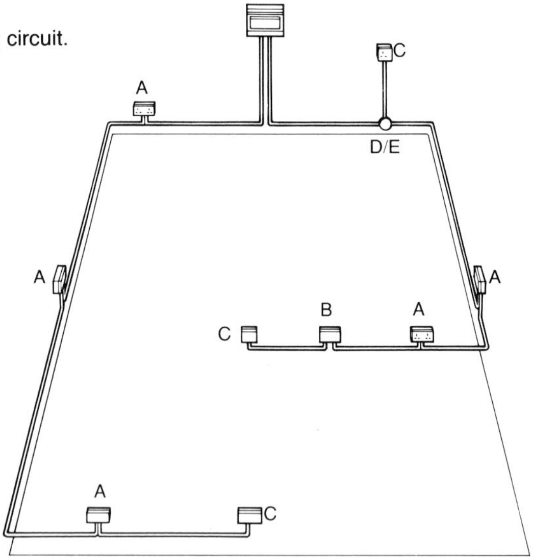

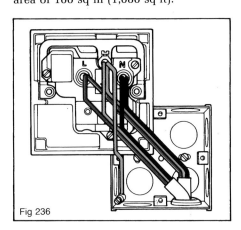

Fig 236

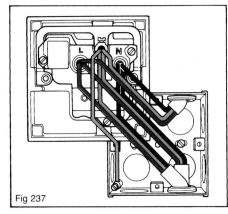

Fig 237

Fig 235   Radial circuit.

## Radial Circuit Check-List

1.  Circuits from 20-amp fuseways wired in 2.5mm² two-core and earth cable can supply an unlimited number of sockets but the floor area of the rooms served must not exceed 20 sq m (215 sq ft). The circuit can be protected by a rewirable or cartridge fuse, or by an MCB.

Fig 236   Socket on ring (A).

Fig 237   Socket feeding spur (B).

Fig 238   Socket on spur (C).

Fig 239   Under-floor junction box on ring (D).

Fig 240   Surface-mounted box on ring (E).

2. Circuits from 30-amp fuseways wired in 4mm$^2$ cable can also supply an unlimited number of sockets, but the floor area of the rooms served by each individual circuit must not exceed 50 sq m (540 sq ft). The circuit must be protected by a cartridge fuse or an MCB, not by a rewirable fuse, to satisfy the wiring regulation requirements.

## Fused Spurs Check-List

1. An FCU on a ring circuit can in theory supply a fixed appliance rated at up to 3.6kW, but overloading could result and a separate circuit is a better solution.
2. Spurs can supply lighting sub-circuits via an FCU containing a 5-amp fuse.

Fig 241 FCU supplying spur cable to appliance.

Fig 242 FCU supplying flex to appliance.

Fig 243 Flex outlet plate with cable from FCU and flex to appliance.

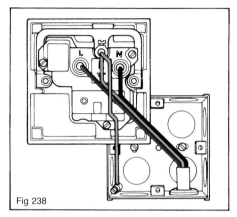

Fig 238

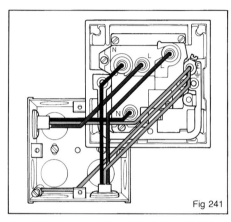

Fig 241

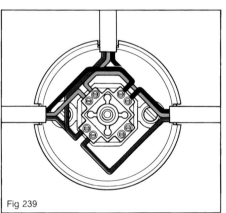

Fig 239

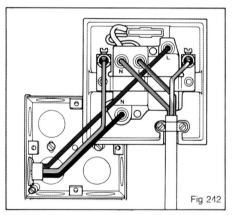

Fig 242

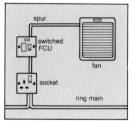

Fig 244 Fused spur from ring circuit socket to supply appliance (l).

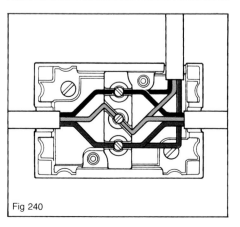

Fig 240

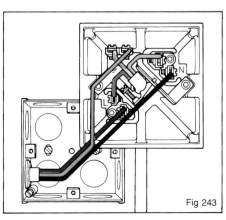

Fig 243

# Lighting Circuits

The range of wiring accessories you will need for your house lighting circuits is smaller than for power circuits. However, since the fittings tend to be a more conspicuous part of each room's decor, how they look is perhaps more important than it is with power circuit accessories. The major manufacturers have taken note of this need, and now offer accessories in a wide range of styles and finishes, making it far easier to match your fittings to the room's style and colour scheme.

As with power circuit accessories, don't forget that you can make big savings by buying in bulk those items you use often.

## Plate-Switches

**Standard single plate-switches**  These are about 85mm square, and are designed to be installed over shallow mounting boxes 16mm deep. Flush metal boxes for plate-switches are often called plaster-depth boxes because the depth is similar to the average plaster thickness on masonry wall. The advantage of this shallow box is that installation merely involves cutting out a recess in the plaster rather than the more laborious removal of solid masonry needed for fitting deeper boxes. However, they do not leave much room for the switch cable inside the box.

**Multi-gang plate-switches**  These are available with two, three, four or six rocker switches on a single face-plate. The two-gang and three-gang types are the same size as a single plate-switch, but the four- and six-gang versions are the same size as a double socket outlet – about 145 x 85mm.

**Architrave switches**  These are slim-line switches designed, as their name implies, to fit on narrow door architraves and in other locations where there is not enough room for a standard plate-switch to be installed. The one-gang version measures about 85 x 32mm, and the two-gang version (which has the two switch rockers mounted one above the other) measures 145 x 32mm. They are mounted over special narrow boxes, 27mm deep for flush metal boxes and 17mm deep for

Fig 245  Plate-switches and lamp-holders.

surface-mounted plastic ones. The flush boxes can also be used as connection boxes for mounting wall lights if the lamp base-plate is too small to fit over a round conduit box (see also page 91, for a description of conduit boxes).

Plate-switches come in two varieties, for one-way and two-way switching. Those intended for one-way switching have just two terminals on the rear of the face-plate, while two-way switches have three terminals marked C, L1 and L2 or L1, L2 and L3 for identification. Usually one-gang switches come in both varieties, but multi-gang switches generally are made only as two-way switches. The switches are usually single-pole only, although double-pole types are also available for use where complete isolation of the light being controlled is necessary.

**Dimmer switches**  These are the same size as plate-switches, but may need deeper mounting boxes – a point to check when buying them as replacements for existing switches. One-gang and two-gang versions are available, usually both the same size as a single plate-switch, although some powerful two-gang dimmers measure 145 x 85mm and need a double width mounting box.

Check the wattage of the lighting to be controlled before buying a dimmer switch. Most will operate only between a minimum and maximum wattage – typically 60 watts to 400 watts for a one-gang dimmer – so you may need to use higher wattage bulbs.

## Junction Boxes

Junction boxes for lighting circuits are always concealed in floor voids, and usually measure 82mm in diameter. They

contain four banks of terminals to cater for loop-in wiring, and have a screw-on cover which can be rotated to reveal cut-outs for various numbers of cable entries round the perimeter of the box.

# Lighting Circuits

## Ceiling Roses

Standard ceiling roses  These measure about 80mm in diameter, and project below ceiling level by 25mm to 30mm. They are designed for surface mounting, although they can be mounted over a circular conduit box if necessary, and have offset knock-outs in the base-plate to allow selection of the most convenient cable entry point, depending on whether the rose is secured to a joist or to a supporting batten.

Most roses nowadays have a bank of terminals which can be wired up on either the loop-in or junction-box methods. In the first case, neutral cores go to one end terminal, circuit and switch live cores to the central terminal and the switch loop core to the other end terminal. With junction-box wiring, only the outer terminals are used. There is a separate earth terminal on the rose base-plate.

The flex to a pendant lamp-holder is connected to the outer terminals, and its cores are looped over supporting hooks to take the strain off the flex terminal connections. Most roses can support fittings weighing up to 5kg (11lb) so long as 1mm² flex is used. For heavier fittings, heavy-duty roses are available.

Plug-in ceiling roses.  The drawback with pendant lamp-holders wired to conventional ceiling roses is that the wiring has to be disconnected if the light needs cleaning, maintaining or repairing. Plug-in roses allow the light to be connected or reconnected easily, by means of a special plug on the pendant flex or within the base of the fitting which locates in a socket on the rose base-plate. Plug-in roses are the same diameter as a standard ceiling rose, but are generally a little deeper.

A similar system is also available for wall lights, consisting of a flush-mounted socket into which the special plug on the light is fitted to connect it to the mains. A blanking plate can be fitted over the box when the socket is not in use.

Many lighting retailers and specialist electrical shops sell conversion kits for existing light fittings and pendants, and a growing number of fittings are now manufactured with an integral plug, enabling fittings to be interchanged easily.

## Lamp-Holders

There are two types of lamp-holder used on lighting circuits.

Pendant lamp-holders  These fit on the end of the flex below ceiling roses supporting lampshades, and contain two terminals to accept the flex cores (metallic lamp-holders also contain a separate earth terminal). All but the cheapest are heat-resistant – a feature worth paying for, since otherwise the lamp-holder may become brittle due to heat rising from the lamp. Lamp-holders for use in bathrooms have a deeper 'skirt' for safety reasons.

Batten lamp-holders  These combine the lamp-holder with a circular base-plate, and are designed for direct mounting on wall and ceiling surfaces. Types for wall-mounting are angled.

## Ceiling Switches

Cord-operated ceiling switches for lighting are similar in style to those for power circuits, if a little smaller. They can be surface-mounted or fitted flush over conduit boxes and are rated at 5 amps or 6 amps. One-way and two-way versions are available, and both can be fitted with a neon indicator light if required.

## Conduit boxes

Many light fittings require a recessed mounting box to house the mains connections, and round conduit boxes are the most common way of providing this. The box has a side or back inlet spigot for the cable and contains fixing lugs at 51mm centres to which the light fitting base (or a cover plate) can be screwed.

# Lighting Circuits

## Lighting Circuit Check-List

1. The circuit is wired from a 5-amp fuseway in the consumer unit.

2. The circuit is wired throughout in 1mm$^2$ two-core and earth PVC-sheathed and insulated cable, except for two-way switching arrangements where special three-core and earth cable is used to link the switches.

3. The circuit can supply a theoretical maximum of twelve lighting points, each notionally rated at 100 watts. In practice each circuit is usually limited to eight lighting points, to allow for the use of more powerful lamps or the installation of multi-lamp fittings without risking overloading the circuit.

4. The circuit is run as a radial circuit, with the cable terminating at the most remote lighting point. Junction boxes can be used to wire in spurs off the main circuit where necessary.

5. There is no limit to the floor area of rooms served by one lighting circuit.

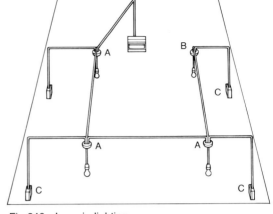

Fig 246   Loop-in lighting circuit.

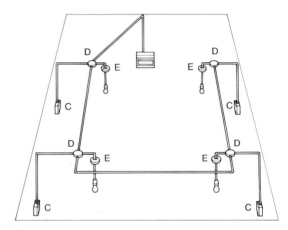

Fig 247   Junction-box lighting circuit.

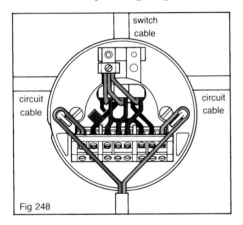

Fig 248

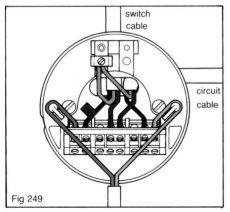

Fig 249

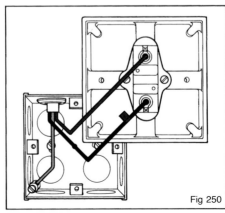

Fig 250

Fig 248   Intermediate loop-in rose (A).

Fig 249   Last loop-in rose (B).

Fig 250   One-way switch (C).

Fig 251   Four-terminal junction box (D).

Fig 252   Rose supplied from junction box (E).

# Lighting Circuits

## Loop-In Wiring Check-List

1.   With loop-in wiring, the circuit cable runs from lighting point to lighting point and switch cables are looped into the rose (or strip connector within conduit boxes).
2.   Cores with black insulation on switch cables must be flagged with red tape.

## Junction-Box Check-List

1.   With junction-box wiring, the circuit cable runs to successive four-terminal junction boxes, terminating at the most remote lighting point.
2.   At each box cables are connected in to supply the individual light fittings and their switches.

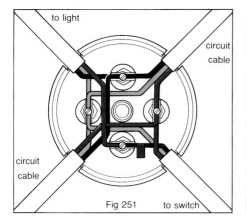

Fig 251

to light

circuit cable

circuit cable

to switch

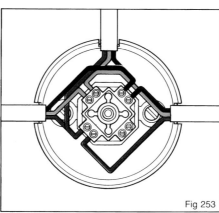

Fig 253

Fig 253   Junction box wired to supply a spur from a lighting circuit (F).

CONDUIT BOXES
Loop-in wiring connections within a conduit box require four strip connectors (see Fig 254).

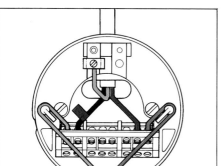

Fig 252

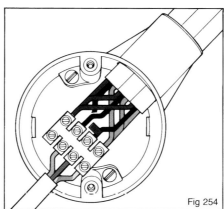

Fig 254

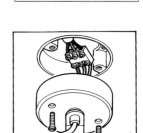

Fig 255   Junction-box wiring connections within a conduit box need only three connectors (H).

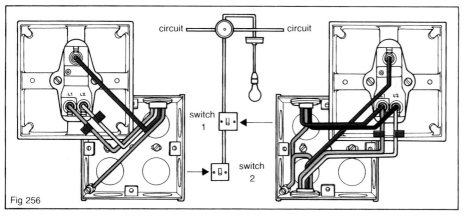

Fig 256

circuit

circuit

switch 1

switch 2

Fig 256   Standard two-way switching arrangement (I).

# Glossary

**Amp (A)** The unit of electric current. Divide an appliance's wattage by mains voltage (240V) to work out how much current it uses.

**Cable** Consists of conductors (or cores) covered with a protective insulating outer sheath, and is used for wiring up individual circuits.

**Circuits** Complete paths round which current flows – along the live conductor to where it is needed, then back to source along the neutral conductor.

**Conduit boxes** Also called BESA boxes, these are round mounting boxes used to contain the connections to light fittings.

**Consumer units** These govern the supply of electricity to all circuits in the house. They contain the system's main on/off switch, and fuses or individual circuit breakers protecting each circuit.

**Double-pole switches** These cut both the live and neutral sides of the circuit, completely isolating the appliance they control from the main supply.

**Earthing** The provision of a continuous conductor on circuits to protect the user from certain electrical faults. Earth conductors are insulated with green and yellow PVC in flex, and where exposed in cables.

**Flex** The short name for flexible cord, this is used to link appliances or pendant lights to the mains.

**Fuses** Protective devices used in circuits and plugs to prevent overloading.

**Gangs** The number of switches or socket outlets in one wiring accessory.

**Grommets** Small rubber washers used in metal mounting boxes to stop the cable chafing on the edges of a knock-out.

**IEE wiring regulations** Guide-lines for safe electrical installation practice. They are issued by the Institution of Electrical Engineers, but do not have any legal force (except in Scotland).

**Insulation** This is present on flex and cable and protects users of electrical equipment from touching live conductors.

**Junction boxes** Used on all circuits to connect in spurs, and on lighting circuits to link cables to roses and switches.

**Knock-outs** Preformed weak spots in mounting boxes, allowing cables to enter.

**Lamps** The correct 'trade' term for light bulbs and tubes.

**Live** This is also called line and describes the cable or flex core carrying current to an accessory or appliance, or any terminal to which this core is connected. Live core insulation is red in cables and brown in flex.

**Loop-in light circuits** These are wired by running cable to each ceiling rose in turn. The switch cable is linked direct to the rose it controls.

**Luminaire support coupler (LSC)** A ceiling-mounted socket to which a pendant light can be connected via a special plug. LSCs are also available for mounting wall lights.

**Miniature circuit breakers** Electro-mechanical switches used instead of circuit fuses in modern consumer units.

**Mounting boxes** Metal or plastic enclosures over which accessory face-plates are fitted.

**Neutral** The cable or flex core carrying current back to its source, or any terminal to which this core is connected. Neutral core insulation is black in cable and blue in flex.

**One-way switches** These control a light from one switch position only. They contain two terminals per gang.

**Radial circuits** Power circuits terminating at the most remote socket outlet on the circuit.

**Residual current circuit breakers (RCCBs)** Protective safety devices fitted to circuits to detect current leakage which could start a fire or cause a shock.

**Ring circuits** Power circuits wired as a continuous loop, both ends being connected to the same live terminals in the consumer unit.

**Single-pole switches** These cut only the live side of the circuit they control. Most light switches are of this type.

**Spurs** Cable branch lines connected to a circuit to supply extra lights or socket outlets.

**Two-way switches** These are used in pairs to allow control of one light from two switch positions. Each gang has three terminals. The switches are linked by special three-core-and-earth cable.

**Volt (V)** The unit of electrical 'pressure' – the potential difference that drives current round a circuit. In most British homes mains voltage is 240V.

**Watt (W)** The unit of power consumed by an appliance or circuit. It is the product of mains voltage x current drawn (in amps). 1000W = 1 kiloWatt.

## USEFUL CONTACTS

Manufacturers

Contactum Ltd
020 8452 6366
www.contactum.co.uk

Cutler Hammer
01404 812131
www.cutler-hammer.eaton.com

Electrium Sales Ltd
(Crabtree, Marbo, Volex and Wylex brands)
01543 455200
www.electrium.co.uk

Legrand Electric
0121 515 0525
www.legrand.co.uk

MEM 250 Ltd
0161 652 1111
www.mem250.com

MK Electric Ltd
01268 563370
www.mkelectric.co.uk

Organisations

Electrical Contractors Association (ECA)
020 7313 4800
www.eca.co.uk

Institution of Electrical Engineers (IEE)
020 7240 1871
www.iee.org.uk

National Inspection Council for Electrical Installation Contracting (NICEIC)
020 7564 2323
www.niceic.org.uk

SELECT (Electrical Contractors Association of Scotland)
0131 445 5577
www.select.org.uk

# Index

# Index